Electrification of Heavy-Duty Construction Vehicles

Synthesis Lectures on Advances in Automotive Technology

Editor
Amir Khajepour, *University of Waterloo*

Electrification of Heavy-Duty Construction Vehicles
Hong Wang, Yanjun Huang, Amir Khajepour, and Chuan Hu
2017

Electrification of Heavy-Duty Construction Vehicles
Hong Wang, Yanjun Huang, Amir Khajepour, and Chuan Hu

ISBN: 978-3-031-00367-7   paperback
ISBN: 978-3-031-01495-6   ebook
ISBN: 978-3-031-00000-3   hardcover

DOI 10.1007/978-3-031-01495-6

A Publication in the Springer series
SYNTHESIS LECTURES ON ADVANCES IN AUTOMOTIVE TECHNOLOGY

Lecture #1
Series Editor: Amir Khajepour, *University of Waterloo*
Series ISSN
ISSN pending.

Electrification of Heavy-Duty Construction Vehicles

Hong Wang, Yanjun Huang, Amir Khajepour, and Chuan Hu
University of Waterloo

SYNTHESIS LECTURES ON ADVANCES IN AUTOMOTIVE TECHNOLOGY #1

ABSTRACT

The number of heavy-duty construction vehicles is increasing significantly with growing urban development causing poor air quality and higher emissions. The electrification of construction vehicles is a way to mitigate the resulting air pollution and emissions. In this book, we consider tracked bulldozers, as an example, to demonstrate the approach and evaluate the benefits of the electrification of construction vehicles. The book is intended for senior undergraduate students, graduate students, and anyone with an interest in the electrification of heavy vehicles.

 The book begins with an introduction to electrification of heavy-duty construction vehicles. The second chapter is focused on the terramechanics and interactions between track and blades with soil. The third chapter presents the architecture and modeling of a series hybrid bulldozer. Finally, the fourth chapter discusses energy management systems for electrified heavy construction vehicles.

KEYWORDS

hybrid electric tracked bulldozer, terramechanics, system modeling, energy management, dynamic programming, model predictive control, hybrid energy storage system

Contents

Preface

This book introduces the electrification of heavy duty construction vehicles. It provides working characteristics and electrified configuration of heavy-duty construction vehicles and also their energy management systems. The book is intended for engineers in construction vehicle companies striving to develop an electrified vehicle and graduate and senior undergraduate students in mechanical and automotive engineering. This book is also accessible to anyone interested in learning about the electrification of heavy-duty construction vehicles. It uses a step-by-step approach using pictures, graphs, tables, and examples so that the reader can easily grasp difficult concepts.

After a short introduction, the terramechanics of a heavy-duty construction vehicles is presented. The architecture of the electrified heavy-duty construction vehicle and modeling of a series hybrid vehicle are introduced. Energy management systems for electrified heavy-duty construction vehicles are discussed and developed. The book ends with conclusions and references.

Hong Wang, Yanjun Huang, Amir Khajepour, and Chuan Hu
November 2017

Acknowledgments

This book would not have been possible without the help of many people. We are particularly grateful to Professor Fengchun Sun and Professor Qiang Song for their support, especially for the modeling and experimental verification. We are also thankful to Shantui Construction Co., Ltd, for experimental data and Tiffany Wong-Zylstra for editing and proofreading the book. We are also thankful to Morgan & Claypool Publishers for providing the opportunity for this book, along with their consistent encouragement and support throughout this project.

Hong Wang, Yanjun Huang, Amir Khajepour, and Chuan Hu
November 2017

CHAPTER 1

Introduction

Off-road applications include equipment used for construction, earthmoving, agriculture, forestry, material handling, recreation, marine purposes, etc. The use of diesel engines in off-road applications is a significant source of nitrogen oxides (NOx) and particulate matter (PM10). In order to prevent global warming, conserve natural resources, and adjust to even more stringent emission regulations, manufacturers of earthmoving equipment are more than ever aware of the importance of producing environmentally friendly machines with significantly improved fuel economy. Although traditional methods have played an important role in reducing energy usage in hydraulic construction machinery, they still result in low fuel efficiency and harmful exhaust gases. New technologies are needed to further reduce fuel consumption and pollutant emissions. With their successful applications in road vehicles, electrified systems are being introduced to traditional construction machineries. Over the past decade, major manufacturers and research institutions worldwide have undertaken several projects to develop hybrid construction vehicles—including the hybrid electric bulldozer, hybrid excavator, and hybrid wheel loader. In this book, a hybrid electric tracked bulldozer (HETB) composed of an engine-generator, ultracapacitor energy storage system, and two driving motors is presented to study electrification of heavy construction vehicles and its impact in improving fuel economy.

Hybrid electric construction machineries are different from road hybrid electric vehicles (HEVs), mainly in three aspects: operation, key components, and reliability (as its components need to be more reliable and robust than road vehicles). As there are differences between hybrid electric construction vehicles and HEVs, their control strategy is also different. In this book, rule-based, dynamic programming and a model predictive controller (MPC) are developed to reduce the fuel consumption of HETBs. The results of three different strategies are compared to show the improvement in fuel efficiency.

In the following, terramechanics analysis of heavy-duty construction vehicles and interactions between track and blades with soil are studied. Using this study, and an architecture for an electrified tracked bulldozer, the modeling of the vehicle is presented. This is followed by a detailed analysis of the energy management of this electrified heavy-duty construction vehicle (including rule-based, dynamic programming and MPC-based energy management), concluding with the hybrid energy source application on this electrified heavy-duty construction vehicle.

CHAPTER 2

Terramechanics of Heavy-duty Construction Vehicle and Interactions Between Track and Blades with Soil

The external loads on bulldozers are the results of interactions between the track and blades with soil. This chapter discusses the mechanics regarding the interaction between the moving device/working part of the bulldozer, and the ground to improve the traction performance and operating efficiency of the bulldozer; and also provides the theoretical foundation for the subsequent driving dynamics analysis of the tracked bulldozer. The interaction forces are also needed in the dynamics and energy management of the electrified vehicle.

During the operation of a tracked bulldozer, the cutting edge of the working part (bulldozing plate) cuts the soil. The energy consumed by the machine, in order to overcome the cutting resistance of the soil, is exerted by the reacting force of the surface soil on the machine [1]. Therefore, during the operation of the machine the cutting resistance of the soil, the support capacity, and horizontal reacting force of the surface soil on the machine, are important factors affecting its performance and efficiency. For many years, the machine's running resistance, work resistance, traction performance, trafficability, etc., were designed and evaluated only using calculation parameters with very large safety factors [2]. For example, in order to simplify the design, the rolling resistance of the bulldozer is often denoted by the product of the rolling resistance coefficient and the weight of the whole machine under certain road conditions [3]. In fact, there are numerous complex factors that can affect the rolling resistance of the moving device for the tracked bulldozer. With the rapid development of science and technology, performance expectations of machines are getting higher, and the effect of many unknown factors (previously covered up by safety coefficients) have gradually become more prominent. The research significance of this chapter is the in-depth study of the interaction between the machine and the ground, and further revealing the interaction forces between the bulldozer and the ground.

2.1 GROUNDING PRESSURE OF TRACK AND ITS EVALUATION INDEX

A tracked bulldozer is driven by the tracks, which plays an important role in the functional application of tracked vehicles, since it bears the weight of the vehicle body and generates traction forces. A tracked vehicle has a ground area dozens or even hundreds of times larger than a wheeled vehicle. Furthermore, the quality of the whole tracked vehicle is fairly evenly distributed on each track shoe through the road wheel. This feature both reduces the pressure of the track on the ground (which can improve the trafficability of the tracked vehicle) and enlarges the contact area between the track vehicle and the ground which can enhance the drivability of tracked vehicles [4].

The ground pressure of the track is the pressure exerted by the track against the ground, which is an important factor of the tracked bulldozer, and has a great impact on the vehicle's driving resistance, traction performance, and trafficability. The ability of the tracked bulldozer to pass the soft ground is usually measured by the Nominal Ground Pressure (NGP), whose value equals to the weight of the bulldozer divided by the ground contact area of the track. NGP is based on the assumption that the track-ground pressure is evenly distributed.

$$NGP = \frac{G}{2bL} \tag{2.1}$$

where,

G the use weight of the bulldozer (N);
b the track width (m);
L the ground contact length of track (m).

In fact, the experimental study of B.G. Schreiner [5] proved that tracked vehicles with the same NGP can have different maneuvering characteristics on the soft ground, i.e., the indicator NGP cannot be used to correctly evaluate maneuverability characteristics of high-speed tracked vehicles on the soft ground. In addition, D. Rowland had studied the pressure changes of 21 kinds of tracked vehicles using sensitive pressure sensors buried in 23 cm of soil, indicating that the ground pressure of tracked machines is not evenly distributed, but related to the quantity and rigidity of the thrust wheels, and the flexibleness of the track, etc. The quantity of the thrust wheels is measured by the ratio of the space between the adjacent thrust wheels S to the track pitch t. The space between the adjacent thrust wheels is called small separation distance when $S/t \leq 2$, and is called large separation distance when $S/t > 2$. The ground pressure will be relatively evenly distributed when there are many thrust wheels, the spacing ratio $S/t \leq 2$, and the track rigidity is large; on the contrary, the ground pressure will be unevenly distributed. The driving speed of the tracked bulldozer is relatively slow, and $S/t \leq 2$, thus NGP under the thrust wheels can be represented by (2.1).

2.2   TRACKED BULLDOZER EXTERNAL DRIVING RESISTANCE

The driving resistance of a tracked bulldozer usually refers to the resistance generated in the traveling of the entire moving device, beginning with the drive wheel. It includes two parts: one is generated by the frictional resistance in the friction pairs of the moving devices, including the friction loss, etc., in the drive bearing, called internal driving resistance; another is generated by the vertical deformation of the soil under the front track when the vehicle is running, called external driving resistance.

2.2.1   COMPACTION RESISTANCE

When a tracked bulldozer travels on soft soil, if the sinking depth is shallow, then the sinking is mainly caused by the vertical plastic deformation of the soil, which includes soil compaction, and the movement of a part of the soil relative to the other part of the soil. The calculation of soil vertical plastic deformation resistance is generally by aid of the principle of energy conservation, which argues that the compaction resistance of the tracked bulldozer is generated by the energy consumption of the vehicle sinking caused by the compaction of the soil body. Therefore, the compaction resistance can be calculated through the energy consumed by the soil compaction, that is, making the work done by the compaction resistance on a certain distance equal to the work done through compacting soil by the bulldozer on the same distance.

As shown in Fig. 2.1, when the tracked bulldozer moves a distance of L, the work P consumed by the vehicle for compacting the soil is:

$$P = \int_0^z 2bLp\,dz \tag{2.2}$$

where,

   p the load acting on the unit supporting surface of the soil (N/m^2);

   z sinking depth (m).

When the tracked bulldozer moves a distance of L, the work to overcome the driving resistance F_c should be equal to the work done by compacting the soil for the vehicle. That is,

$$F_c \bullet L = P = 2bL \int_0^z p\,dz$$

$$F_c = 2b \int_0^z p\,dz \tag{2.3}$$

Applying the Biluliya's pressure subsidence formula [165]:

$$p = kz^n \tag{2.4}$$

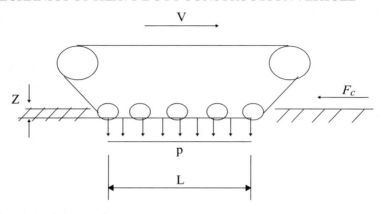

Figure 2.1: Schematic diagram for the calculation of the compaction resistance.

where,

> k deformation modulus of the soil (KN/m^{n+2}), $k = \frac{k_c}{b} + k_\varphi$;
>
> k_c coefficient of cohesion of the soil (KN/m^{n+1});
>
> k_φ internal friction coefficient of the soil (KN/m^{n+2}); and
>
> n deformation index of the soil, zero dimension.

Then,

$$F_c = 2b \int_0^z kz^n \, dz = 2bk \frac{z^{n+1}}{n+1} \tag{2.5}$$

According to (2.1), we have

$$p = \frac{G}{2bL} \tag{2.6}$$

Thus,

$$z = \left(\frac{p}{k}\right)^{\frac{1}{n}}$$

$$F_c = \frac{2b}{(n+1)k^{\frac{1}{n}}} \left(\frac{G}{2bL}\right)^{\frac{n+1}{n}} \tag{2.7}$$

Equation (2.6) clearly shows that the external driving resistance of the tracked vehicle depends on the vehicle parameters and the physical and mechanical properties of the soil. From the above equation we can make the following three conclusions:

1. Reducing the ground pressure can reduce the vehicle's driving resistance, that is, increasing the ground contact area or reducing the vehicle weight will reduce the driving resistance.

2. When the ground contact area is constant, the long and narrow track has smaller driving resistance than the short and wide track.

3. The external driving resistance of the vehicle is highly related to the physical and mechanical properties of the soil. In the dry and dense soil, the vehicle driving resistance is small, however when the vehicle works in soil with less carrying capacity (such as marshes), once the ground pressure is greater than the ultimate load capacity of the soil, the soil body will be destroyed, and the vehicle cannot move.

2.2.2   BULLDOZING RESISTANCE

When the bulldozer is running on soft ground, the bulldozer's subsidence is considerable. In addition to the compaction resistance, due to the formation of the wheel cutting of the compacted soil, the moving device of the track is also subjected to the effect of bulldozing resistance. When the tracked bulldozer is running, there is a soil uplift at the front of the track. The resistance of the soil uplift at the front of the moving device is commonly called the bulldozing resistance. The principle of formation of the bulging resistance is similar to that of the retaining wall, and can be calculated using the passive earth pressure theory on the retaining wall.

The force situation of the bulldozer track shoe is shown in Fig. 2.2, where the track shoe is subject to a passive earth pressure F_b.

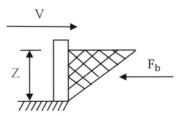

Figure 2.2: The schematic diagram of the bulldozing resistance.

According to the Langkin earth pressure formula [165], the following equation can be used to approximately express the passive earth pressure of the tracked bulldozer:

$$F_b = \gamma Z^2 b K_\gamma + 2bZcK_{pc} \tag{2.8}$$

$$K_\gamma = \left(\frac{2N_\gamma}{\tan \varphi} + 1\right)\cos^2 \varphi$$

$$K_{pc} = (N_c - \tan \varphi)\cos^2 \varphi$$

where

N_γ, N_c bearing capacity factor in the whole shear failure;

γ weight of the soil (N/m³).

In summary, the bulldozing resistance of the bulldozer moving device is related to the soil parameters, and is also in direct proportion to the track width. Reducing the track width can reduce the bulldozing resistance.

2.2.3   OPERATING RESISTANCE

The operating resistance of the bulldozer F_T mainly includes [5]: tangential cutting resistance F_1, advancing resistance of the soil before the blade F_2, friction resistance between the cutting edge and soil F_3, horizontal component force of the friction resistance when the soil crumb is rising along the blade F_4.

$$F_T = F_1 + F_2 + F_3 + F_4 \tag{2.9}$$

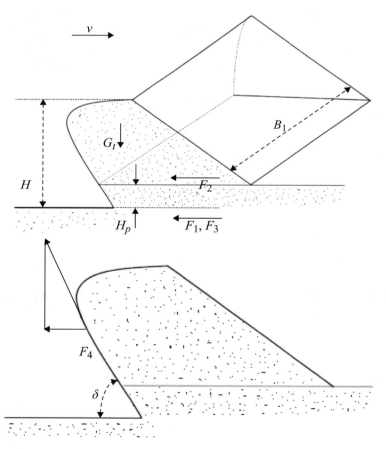

Figure 2.3: Schematic diagram of the operating resistance of the bulldozer.

(1) Tangential cutting resistance F_1

$$F_1 = 10^6 B_1 h_p k_b \qquad (2.10)$$

where,

B_1 width of the bulldozer shovel (m);
h_p cutting depth of the bulldozer shovel (m);
k_b cutting ratio resistance (MPa).

(2) Advancing resistance of the soil before the blade F_2

$$F_2 = G_t \mu_1 \cos \alpha = \frac{V \gamma \mu_1 \cos \alpha}{k_s} \qquad (2.11)$$

$$V = \frac{B_1 \left(H - h_p \right)^2 k_m}{2 \tan \alpha_0} \qquad (2.12)$$

where,

G_t the gravity of the mound before the bulldozing plate (N);
V the volume of the mound before the bulldozing plate (m^3);
k_s loose coefficient of the soil;
k_m filling coefficient of the soil;
H height of the shovel blade (m);
μ_1 friction coefficient of between soil and soil;
γ weight of the soil (N/m^3);
α gradient (°);
α_0 repose angle of the soil (°).

(3) Friction resistance between cutting edge and soil F_3

$$F_3 = 10^6 B_1 X \mu_2 k_y \qquad (2.13)$$

where

k_y ratio resistance of the cutting edge compacting the soil after the cutting edge is worn (MPa);
X ground contact length of the cutting edge after abrasion (m);
μ_2 friction coefficient between the soil and steel.

(4) Horizontal component force of the friction resistance when the soil crumb is rising along the blade F_4

$$F_4 = G_t \mu_2 (\cos \delta)^2 \cos \alpha \qquad (2.14)$$

where,

δ cutting angle of the bulldozer shovel (°).

2.2.4   OTHER RESISTANCES

(1) Grade resistance F_i

$$F_i = G \sin \alpha \qquad (2.15)$$

where,

G working weight of bulldozer (N);

α gradient (°).

(2) Air resistance F_w

$$F_w = \frac{C_D A}{21.15} v^2 \qquad (2.16)$$

where

C_D air resistance coefficient;

A windward area (m^2);

v vehicle relative driving speed (km/h).

(3) Acceleration resistance F_j

$$F_j = \delta m \frac{du}{dt} \qquad (2.17)$$

where,

δ conversion factor of the vehicle rotation mass;

m mass of the whole vehicle (kg);

$\frac{du}{dt}$ driving acceleration (m/s^2).

Since the velocity and acceleration of the bulldozer are very low during operation, the air resistance and acceleration resistance can be neglected.

2.3   TRACKED BULLDOZER TANGENT TRACTION FORCE

Under the action of the vehicle, the horizontal thrust generated by the deformation of the ground is called tangential traction or soil thrust. When the vehicle is running, the drive wheel will rotate or drive the track to rotate under the action of the driving torque, and this motion will be prevented by a reactive force generated by the ground, which is called the tangential traction. When the vehicle runs on the hard road, the tangential traction is mainly generated by the friction between the moving devices and the ground. When the vehicle runs on the soft ground, the tangential traction is mainly the soil thrust caused by the shearing strength when the soil has the shear deformation under the action of the tracks. Part of the soil thrust is consumed to overcome the running resistance, and the rest is used for machine work and the vehicle accelerating, climbing, or traction loads.

2.3.1   TRACTION AND SLIP RATIO

First look at the relationship between the soil shear deformation on the track-ground contact area and the track-slip rotation.

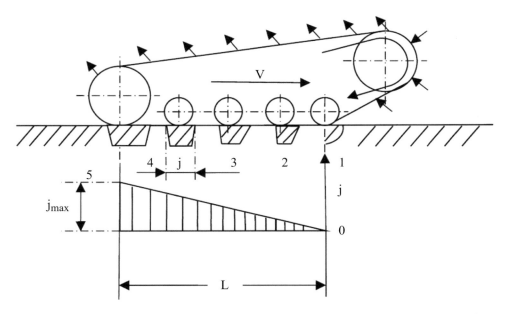

Figure 2.4: Soil shear displacement along the track-ground contact length of the bulldozer.

On point 1 (under the first thrust wheel), the track just makes contact with the ground, thus the shear deformation is zero. The shear displacements on points 2–5 are larger than that on point 1, since the shear deformation has lasted for an amount of time with respect to the ground. The sum of the shear displacement j accumulates along the ground contact length, and reaches a maximum at the back of the contact surface.

In order to quantitatively investigate the generation of the shear displacement under the track, firstly we need to define the slip rotation rate of the track i:

$$i = 1 - \frac{v}{rw} = 1 - \frac{v}{v_T} = \frac{v^T - v}{v_T} = \frac{v_j}{v_T} \tag{2.18}$$

where

- i track slip rotation rate;
- v real velocity of the vehicle (track) (m/s);
- v_T theoretical velocity of the vehicle (track), determined by the angular velocity and pitch radius of the chain wheel (m/s);
- w angular velocity of the chain wheel (rad/s);
- r pitch radius of the chain wheel (m);
- v_j slip rotation velocity of the track with respect to the ground (rad/s).

When the bulldozer is in slip rotation, the direction of v_j is opposite to the driving direction. Conversely, when the bulldozer is in slippage, the direction of v_j is the same as the driving direction. Since the track is unstretchable, the slip rotation velocity of every point on the track which makes contact with the ground is identical.

The shear displacement of the point with a distance x from the front-end of the track contact area is:

$$j = v_j t \tag{2.19}$$

where t is the contacting time of this point with the ground, whose value is $t = x/v_T$. Substitute it into the above equation we have:

$$j = \frac{v_j x}{v_T} = ix \tag{2.20}$$

which indicates that the shear displacement of the track on the contact surface increases linearly from front to back.

According to the Garossion shear stress-displacement relationship [165]:

$$\tau = (c + \sigma \tan \varphi)\left(1 - e^{-\frac{j}{k}}\right) \tag{2.21}$$

where,

τ shear stress;

c soil cohesion force;

σ normal stress on the shearing surface with an assumption
that the track-ground pressure is uniformly distributed,
$\sigma = \frac{G}{2bL} = p$;

φ internal friction angle of the soil;

K horizontal shear deformation modulus, which is a required
deformation value to achieve the maximum shear stress.
This value is determined by the soil compressibility.
Reece points out that K value of the loose sand is 0.5 cm,
K value of the frictionless clay under the maximum compression
is about 0.6 cm.

The total soil thrust of the tracked bulldozer is:

$$F = 2b \int_0^L \tau dx \tag{2.22}$$

Substitute (2.18) and (2.19) into the above equation, we have:

$$
\begin{aligned}
F &= 2b \int_0^L \tau dx \\
&= 2b \int_0^L (c + \sigma \tan \varphi) \left(1 - e^{-\frac{j}{k}}\right) dx \\
&= 2b \int_0^L \left(c + \frac{G}{2bL} \tan \varphi\right) \left(1 - e^{-\frac{ix}{K}}\right) dx \\
&= (2bLc + G \tan \phi) \left[1 - \frac{K}{iL} \left(1 - e^{-\frac{iL}{K}}\right)\right]
\end{aligned}
\tag{2.23}
$$

When the track completed slips, i.e., $i = 1$, $e^{-\frac{iL}{K}} \approx 0$, thus $F_{\max}$ can be calculated as:

$$F_{\max} = 2bLc + G \tan \varphi \tag{2.24}$$

The track of the tracked bulldozer is mounted with the grouser with the height of h, when the grouser is not very low, and does not have a very dense interval, the side effect of the grouser should be considered. This is because the lateral shears' force along the two sides of the track shoe increases the tangential traction force. Figure 2.5 is the structure diagram of the track grouser of the SD24-5 bulldozer in the testing site.

Figure 2.5: Structure diagram of the track of the SD24-5 bulldozer.

Considering the additional traction force of the grouser ΔF, we have

$$\Delta F = \left\{ 4hLc + 0.64G \tan\varphi \left[\frac{h}{b} \operatorname{arc cot}\left(\frac{h}{b}\right) \right] \right\} \left(1 + \frac{K}{iL} e^{-\frac{iL}{K}} - \frac{K}{iL} \right) \qquad (2.25)$$

where h is the height of the grouser. Thus, the adhesive force $F_{\max}$ can be expressed more completely as

$$F_{\max} = 2bLc + G \tan\varphi + 4hLc + 0.64G \tan\varphi \left[\frac{h}{b} \operatorname{arc cot}\left(\frac{h}{b}\right) \right]$$
$$= 2bLc \left(1 + \frac{2h}{b} \right) + G \tan\varphi \left\{ 1 + 0.64 \left[\frac{h}{b} \operatorname{arc cot}\left(\frac{h}{b}\right) \right] \right\} \qquad (2.26)$$

The relationship of the traction force F and the slip rotation rate i can be formulated as

$$F = F_{\max} \left[1 - \frac{K}{iL} \left(1 - e^{-\frac{iL}{K}} \right) \right] \qquad (2.27)$$

2.3.2   CASE STUDY

Since when the bulldozer is under the general operation, the soil has been loosened by the ripper, it is reasonable to take the sandy soil, for example, to calculate the slip rotation curve of the track by combing the size parameters of the SD24-5 bulldozer. The table below shows the related parameters of the SD24-5 bulldozer size and the soil.

From Equations (2.27) and (2.28), we can calculate the slip rotation curve of a SD24-5 bulldozer on the sandy soil as shown in Fig. 2.6.

From Fig. 2.6, we can see that when the track completely slips, i.e., $i = 1$, the traction force reaches the maximum, approximately 213 KN. The difference between the traction force

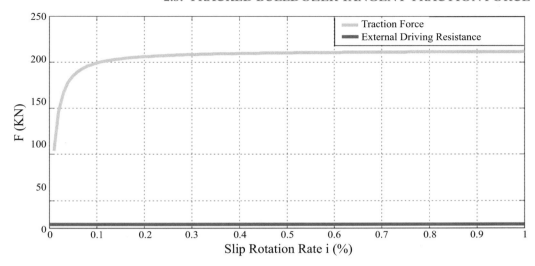

Figure 2.6: Slip rotation curve.

and the external driving resistance is the drawbar traction force, which is used for the bulldozer work, acceleration, climbing, or pulling loads. When the driving force of the bulldozer does not exceed the maximum adhesive force $F_{\max}$, it can be regarded that the driving force can be completely generated; however, once the driving force exceeds the maximum adhesive force, the track will slip.

In the following, we will analyze, under typical bulldozer working conditions, whether the total resistance borne by the track exceeds the maximum driving force $F_{\max}$. Figure 2.7 shows the typical working conditions of the bulldozer (40 m soil-cutting operations) according to the JB/T1666-1997 "*Experimental methods of the tracked bulldozer.*"

In Fig. 2.7, the abscissa is the time, with a total duration of 118 seconds. The ordinate from top to bottom is respectively, the velocity of the bulldozer (km/h), soil-cutting depth (m), climbing slope (°), and steering signal. The bulldozer speeds up in 0–2 s, cuts the soil in 2–37 s, bulldozes the soil in 27–66 s; offloads the soil in 66 s; reverses and returns to the original place in 66–118 s, and then prepares for the next-round soil-cutting operation.

Figure 2.8 shows the force diagram of the unilateral track in this typical working condition.

From Fig. 2.8, we can see the maximum one-sided track resistance is approximately 100 KN. Since the tracked bulldozer is in straight driving condition, the total force of the two side tracks is equal to the sum of two one-sided track resistances—therefore, the maximum total resistance of the tracks is 200 KN. To sum up, in the typical working conditions, the total resistance of the tracks does not exceed the maximum adhesive force limit $F_{\max}$.

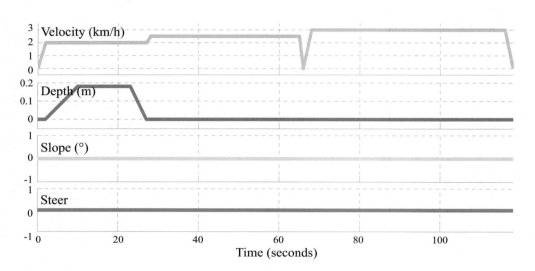

Figure 2.7: Typical working conditions of the bulldozer.

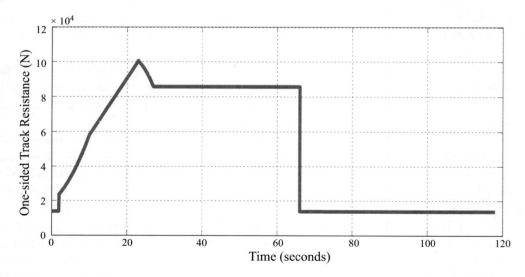

Figure 2.8: One-sided track resistance in 40 m soil-cutting operation working condition.

Table 2.1: Related parameters of the vehicle and soil

Name	Unit	Value
H (Shovel blade)	m	0.07
L (Track-ground contact length)	m	3.05
B (Track width)	m	0.61
G (Working weight of the bulldozer)	KN	280
c (Cohesive force of the soil)	KPa	13.79
φ (Internal friction angle of the soil)	^{o}	28
n (Deformation index of the soil)	$null$	0.3
k_c (coefficient of cohesion of the soil)	KN/m^{n+1}	2.79
k_φ (Internal friction coefficient of the soil)	KN/m^{n+2}	141.11
A (Windward area)	m^2	3.7
γ (Unit weight of the soil)	N/m^3	17700
K (Horizontal shear deformation modulus)	m	0.02

2.4   DISTRIBUTION OF GROUNDING PRESSURE IN A TRACKED BULLDOZER

For the tracked bulldozer, the maximum track-ground pressure occurs in the working conditions. Thus, the research on the track-ground pressure in working conditions is significant. In order to simplify the mathematical model, some assumptions are made as follows:

(1) $S/t \leq 2$, where S is the space between the thrust wheels of the tracked bulldozer, and t is the track pitch, thus there will not be large wave peaks for the ground pressure under the thrust wheels. It can be assumed that the track-ground pressure has a linear distribution along the ground contact length;

(2) The distribution of the track-ground pressure is affected by the center-of-gravity position, working resistance, machine weight, surface slope, and velocity, etc.;

(3) In the working conditions, the effects of the track slip rotation sinking on the distribution of the track-ground pressure and the angle between the track and the ground (which is very small, $\alpha \leq 6°$) can be neglected;

(4) The machine center of the tracked bulldozer has a very small lateral deviation, which can be neglected;

(5) The working load of the tracked bulldozer is uniformly distributed on the bulldozing plate or ripper, where the lateral load is very small, and can be neglected.

2.4.1   EXTERNAL LOAD ANALYSIS OF THE TRACKED BULLDOZER-SOIL SYSTEM

(1) Resultant force of the normal counter-forces on the track device ground contact area F_Q.

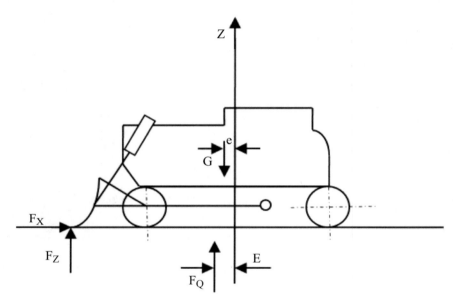

Figure 2.9: The working conditions of the tracked bulldozer.

Figure 2.9 regards the bulldozer as the free body. According to the equilibrium force equation $\sum F_z = 0$, we can obtain the normal counter-force on the track device ground contact area as

$$F_Q = G - F_Z \tag{2.28}$$

where F_Z is the normal cutting resistance of the soil body exerting on the shovel blade or ripper, which is determined by the soil property, cutting depth, pile mass, blade geometry parameters, and dimension parameters, and is positive when the direction is upward.

(2) Moment of reaction on the track-ground contact area.

Considering the free body diagram of the bulldozer and according to the moment equilibrium equation $\sum M = 0$, we have

$$M_{QO} = eG - M_{XO} - M_{ZO} \tag{2.29}$$

where

M_{QO} Moment of reaction with respect to the geometry center of the track-ground contact area, which is positive when the direction is clockwise, otherwise, it's negative;

e Longitudinal offset of the machine center of gravity, which is positive when the offset is forward;

M_{XO} Longitudinal moment caused by the tangential resistance F_X on the shovel blade or ripper exerted by the soil body with respect to the geometry center of the track-ground contact area, which is positive when the direction is clockwise, otherwise, it's negative;

M_{ZO} Longitudinal moment caused by the normal resistance F_Z on the shovel blade or ripper exerted by the soil body with respect to the geometry center of the track-ground contact area, which is positive when the direction is clockwise, otherwise, it's negative.

(3) The longitudinal offset E of the normal counter-force on the track-ground contact area.

With knowing the moment of reaction with respect to the geometry center of the track-ground contact area,

$$M_{QO} = F_Q E \tag{2.30}$$

we have the longitudinal offset of the normal counter-force on the track-ground contact area

$$E = \frac{eG - M_{XO} - M_{ZO}}{F_Q} \tag{2.31}$$

If E is positive, then we know the soil normal counter-force is forward, otherwise, it is backward.

(4) Distribution function of the track-ground pressure.

The track-ground pressure has a trapezoid or triangle distribution along the ground contact length L. According to the related mechanics theory, we can deduct the distribution function of the track-ground pressure

$$\begin{aligned} p &= \frac{F_Q}{2bL} + \frac{M_{QO}}{J}x \\ &= \frac{F_Q}{2bL} + \frac{eG}{J}x - \frac{M_{XO}}{J}x - \frac{M_{ZO}}{J}x \end{aligned} \tag{2.32}$$

where

J Inertia moment of the track-ground contact area, $J = \frac{1}{6}bL^3$;

x Distance from any track-ground contact point to the track-ground contact geometric center axis, which is positive when it is on the left of the original point, otherwise, it is negative, i.e., the vehicle's forward direction.

When the bulldozer stops working, i.e., $F_X = F_Z = M_{XO} = M_{ZO} = 0, F_Q = G$, we have

$$p = \frac{G}{2bL}\left(1 + \frac{12e}{L^2}x\right) \qquad (2.33)$$

Furthermore, since the basic principle of arranging the center-of-gravity position of the tracked bulldozer is to guarantee that the ground pressure in the main working conditions is evenly distributed, and its degree of non-uniformity in other working conditions should not be too large, also, the bulldozer has longitudinal and lateral stability. Thus, it is reasonable to assume that the degree of deviation of the center of gravity is very small. If we neglect the longitudinal offset of the center of gravity position, we can obtain the track ground pressure distribution as

$$p = \frac{G}{2bL} \qquad (2.34)$$

CHAPTER 3

Architecture and Modeling of Electrified Heavy-duty Construction Vehicles

Electrification possibilities of powertrains and auxiliary devices in heavy-duty vehicles are growing because of advances in related technologies and products [1]. An electrified powertrain consists of fewer parts, especially moving ones, which enhance the powertrain's reliability due to fewer failures and maintenance issues.

Hybrid powertrain designs rely on a diesel engine and electric motor(s) to provide power either in a sole or assistance mode to the vehicle. Despite configuration type, as analyzed below, the overall fuel efficiency and the emission discharge can be significantly improved compared to conventional vehicles by the vehicle electrification. In addition, the auxiliary devices including saws, balers, and dozers in heavy-duty vehicles used in construction, commercial, and agriculture, can also be electrified to increase productivity by enabling high response speed and accuracy. The electrification of the auxiliary power devices is extensively studied in [2, 3].

In passenger vehicles, the configuration of the hybrid electric vehicles usually is categorized into three classes: series, parallel, and power-split. Thus, the same categories are adopted here to classify heavy-duty hybrid vehicles.

Series hybrid powertrain

Series hybrid powertrain is the simplest configuration, where only an electric motor instead of a diesel engine drives the wheels. The motor is powered by either the energy buffer (e.g., ultracapacitor or battery) or the engine via a generator. The energy buffer can be recharged by the engine/generator or regenerated braking when available. A supervisory energy management controller determines the power from the energy buffer and engine simultaneously to pursue the maximum overall powertrain efficiency. The major benefit of this configuration is separating the engine from the wheels. As a result, the engine is able to be controlled to mainly operate in its high-efficiency region. Figure 3.1 [4] shows a series hybrid powertrain of a bulldozer from Shantui Construction Machinery Co., Ltd, where ultracapacitors are used as the energy buffer to store the extra energy from the engine or recaptured energy during the braking period.

The Volvo wheel loader (e.g., LX1) [5], the John Deere wheel loader 644 K [6], and Oerlikon excavator [7] also belong to the series hybrid configuration. Compared to the only

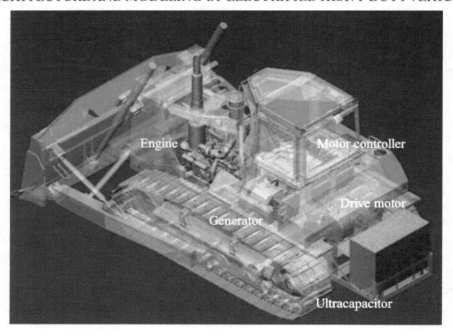

Figure 3.1: Powertrain of a series hybrid bulldozer.

diesel engine driven model, the new powertrain results in up to 30% fuel savings and even more when a construction machine experiences a large number of stop/go driving conditions. For example, a wheel loader usually spends 40% of the working time idling [8].

Parallel hybrid powertrain

In construction vehicles with the parallel hybrid powertrain, the engine and the electric motor work together to power the wheels via a mechanical transmission [12]. Because of the addition of an electric motor and a corresponding energy storage system, the engine can be downsized. This is the main advantage of the parallel hybrid configuration. Since the engine is coupled with wheels, compared to the series configuration, the efficiency benefits in frequent stop-and-go scenarios are limited. However, it yields efficiency benefits as the need to convert mechanical energy to electricity and back to mechanical energy is eliminated. The hybrid wheel loader L220F from Volvo and the 12MTX Hybrid wheeled excavator from Mecalac Ahlmann [12] are examples of the parallel hybrid construction vehicles.

Power-split hybrid powertrain

In passenger vehicles, this configuration combines the advantages of both the series and parallel hybrid ones via using a planetary gear set. However, due to its complex structure and large nominal power of construction vehicles, very few commercial construction vehicles use this configuration.

Electric-driven powertrain

Unlike the passenger cars, another commonly used electrified configuration in heavy-duty vehicles is the electric-driven powertrain, which is similar to the series hybrid, but it is not the hybrid one because the diesel engine is the sole power source and does not have regenerative braking ability. The main generator is connected to the diesel engine for electricity, which is transferred to the electric motor to drive the wheels. In this process, the braking energy is consumed by braking resistors. This configuration is commonly used in heavy-duty construction or mining trucks. Due to the hundreds of tons vehicle mass, the braking energy is always too large to be captured and stored. The Caterpillar D7E bulldozer and DE400 from Xuzhou Construction Machinery Group are in this category. While it does not have the ability to recover the braking energy, this design can still provide some energy efficiency improvements as the engine can always work in the high-efficiency region.

3.1   MODELING OF SERIES HYBRID BULLDOZER

The series hybrid powertrain, as indicated in Fig. 3.2, consists of a diesel engine, a permanent magnet generator, and a pair of motor drive systems and tracks. The values of the main parameters are presented in Table 3.1. The hybrid powertrain utilizes an integrated controller to independently manipulate the motors on both sides, and the electric energy from the generator and auxiliary power supply are transferred into the mechanical energy to drive the bulldozer. According to the working principle and the experiment of various parts of the powertrain, a mathematical model of the hybrid powertrain is developed.

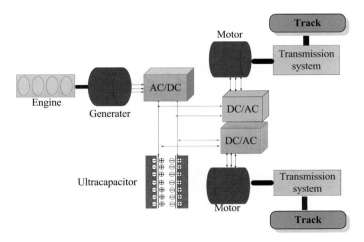

Figure 3.2: Layout of the hybrid electric tracked bulldozer.

Table 3.1: Main vehicle parameters

Component	Parameters	Quantity
Diesel	Maximum power	172kW/1800 rpm
Engine	Maximum torque	1087Nm/1300 rpm
Generator	Maximum power	180 kW
	Maximum torque	1,010 Nm
	Rated speed	1,700 rpm
Motor	Maximum power	105 kW
	Maximum torque	800 Nm
	Rated speed	1,430 rpm
Vehicle	Curb weight	28,000 kg
	Track width	0.61 m
	Track length	2.05 m
	Drive wheel radius	0.46831 m

3.1.1   ENGINE-GENERATOR MODEL

To derive an accurate engine model, the engine universal characteristic curves shown in Fig. 3.3 obtained from experimental data are employed to model the engine.

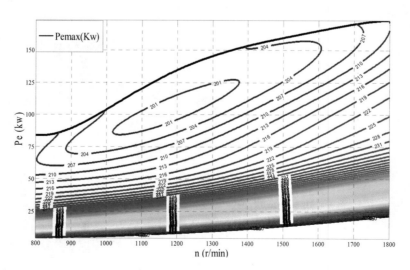

Figure 3.3: Diesel engine universal characteristic curve.

The fuel consumption depends on the throttle position and engine speed. Engine output torque is represented by:

$$T_e^D = T_e - J_e \frac{dw_e}{dt} \tag{3.1}$$

$$T_e = \alpha * T_{e\_\max}(n_e) \tag{3.2}$$

where T_e^D is the dynamic engine output torque; T_e denotes the steady-state engine flywheel output torque; α refers to the throttle position; $T_{e\_\max}(n_e)$ represents the maximum torque at the engine speed n_e; and J_e indicates the rotary inertia of the engine rotating parts.

The fuel consumption b_e (g/kWh) is modeled as a function of T_e and n_e:

$$b_e = f(T_e, n_e) = \sum_{j=0}^{s} \sum_{i=0}^{j} A_k T_e^i n_e^{j-1} \quad (i, j = 0, 1, 2, \ldots, s) \tag{3.3}$$

where s is the order number, and A_k denotes the polynomial coefficient $k = (j^2 + j + 2 * i)/2$.

The generator provides electricity to DC BUS, and it can be modeled as:

$$J_g \frac{dn_g}{dt} = T_g - T_e^D \tag{3.4}$$

where T_g is the torque, n_g denotes its speed, and J_g is the rotary inertia.

The correlation between the torque, speed, the output voltage, and current is represented by:

$$\frac{U_g \cdot I_g}{1000 \cdot \eta_g} = \frac{T_g \cdot n_g}{9549} \tag{3.5}$$

where U_g is the voltage, I_g is output current; η_g is the efficiency.

3.1.2 DRIVE MOTOR MODEL

In this application, two permanent magnet motors are used with a 500/800 N.m torque, a 75/105 Kw power, and a 1430/6000 rpm speed. The controller determines their output torques. Therefore, the motor efficiency map shown in Fig. 3.3 is obtained based on the experimental data.

Given the response time of the motor drive system, a first order link is introduced:

$$T_m = \begin{cases} \dfrac{T_{ref}}{\tau s + 1} & T_{ref} \leq T_{\max}(n) \\[2mm] \dfrac{T_{\max}(n)}{\tau s + 1} & T_{ref} > T_{\max}(n) \end{cases} \tag{3.6}$$

where T_{ref} is the target torque, and T_m means the output torque. n is the motor speed, and $T_{\max}(n)$ is maximum torque at speed n. τ is the response time.

The dynamic equation of the motor is represented by:

$$J_m \frac{dn}{dt} \frac{2\pi}{60} = T_m - T_{load} \tag{3.7}$$

where J_m is the motor inertia, and T_{load} is load torque.

Similarly to (3.5), the correlation between the input DC voltage, the current of the motor controller, the shaft output torque, and the speed is:

$$\begin{cases} \dfrac{U \cdot I \cdot \eta_d}{1000} = \dfrac{T_m \cdot n}{9549} & (T_m > 0) \\[3mm] \dfrac{U \cdot I}{1000 \cdot \eta_b} = \dfrac{T_m \cdot n}{9549} & (T_m < 0) \end{cases} \tag{3.8}$$

where U and I is the input DC voltage and DC current, respectively. η_d is motor efficiency when driving, and η_b is motor efficiency when braking.

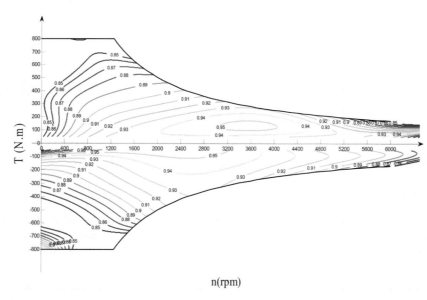

Figure 3.4: Motor drive system efficiency.

3.1.3   ULTRACAPACITOR MODEL

An ultracapacitor pack consists of several cells in both series and parallel patterns, which can be modeled as a series combination of a capacitance and a resistance. Therefore, the simple equivalent RC model is adopted.

$$P_{uc}(t) = V_L(t) \cdot I_{cap}(t) \tag{3.9}$$

$$V_{cap}(t) = -\frac{1}{C} I_{cap}(t) \tag{3.10}$$

$$SOE(t) = \frac{E(t)}{E_{cap}} = \frac{\frac{1}{2} C V_{cap}(t)^2}{\frac{1}{2} C V_{\max}^2} = \frac{V_{cap}(t)^2}{V_{\max}^2} \tag{3.11}$$

where V_L is the terminal voltage, and V_{cap} is the open-circuit voltage. C is the equivalent capacitance; I_0 is the ideal output current; I_{cap} is the actual output current. $V_{\max}$ is the maximum voltage; $E(t)$ is the energy capacity; SOE denotes the state of energy; E_{cap} represents the maximum energy capacity.

The correlation representing the differential of SOE, the maximum energy capacity, and the power is shown in (3.11). Since the problem is modeled by the power balance equations, selecting SOE as the state is more natural. The dynamic equation of the SOE is:

$$SOE(t) = \begin{cases} -\dfrac{1}{\eta_{cap}} \dfrac{P_{uc}(t)}{E_{cap}} & \text{if } P_{uc}(t) \geq 0 \text{ (discharge)} \\[4mm] -\eta_{cap} \dfrac{P_{uc}(t)}{E_{cap}} & \text{if } P_{uc}(t) < 0 \text{ (charge)} \end{cases} \tag{3.12}$$

where η_{cap} is the efficiency.

The main parameters of the UC pack are shown in Table 3.2.

Table 3.2: Ultracapacitor pack parameters

Description	Quantity	Unit
Nominal capacity	2.4	[F]
Nominal voltage	600	[V]
Peak cell discharging current	100	[A]
Weight	260	[kg]
Volume	600	[mm]
Number of cycles	>50,000	—

3.1.4 VEHICLE DYNAMIC MODEL

Different from the road vehicle dynamics (considering the rolling, slope, acceleration, and aerodynamic drag resistance), the two main resistances of the bulldozer are the operating resistance and the external travel resistance. The acceleration resistance and the aerodynamic drag resistance are negligible. Thus, the power requirement can be calculated by:

$$P_{req} = v * (F_E + F_T)$$

$$= v * \left[\begin{array}{l} \frac{2b}{(n+1)k^{\frac{1}{n}}} \left(\frac{G}{2bL'}\right)^{\frac{n+1}{n}} + \gamma Z^2 b K_\gamma + 2bZcK_{pc} \\ + 10^6 B_1 h_p k_b + \frac{V\gamma\mu_1 \cos\theta}{k_s} + 10^6 B_1 X\mu_2 k_y + G_t\mu_2 \cos\delta^2 \cos\theta \end{array} \right] \quad (3.13)$$

where,

$$Z = \left(\frac{G}{2bL'k}\right)^{\frac{1}{n}}$$

$$K_\gamma = \left(\frac{2N_\gamma}{\tan\psi} + 1\right) \cos^2\psi$$

$$K_{pc} = (N_c - \tan\psi) \cos^2\psi$$

$$V = \frac{B_1 (H - h_p)^2 k_m}{2\tan\alpha_0} \quad (3.14)$$

and v denotes the bulldozer speed (m/s); F_E refers to the external travel resistance (N); F_T means the operating resistance (N); b and L represent the track width (m) and length (m), respectively. c indicates the soil cohesion coefficient (kPa); G is the vehicle's weight (N); Ψ is the soil internal friction angle (°); n is the soil deformation index; k is the soil deformation modulus (KN/m^{n+2}); Z is the track's amount of sinkage (m); γ is the unit weight (N/m^3); and N_γ and N_c are the soil Terzaghi coefficients of the bearing capacity. k_b is the cutting resistance per unit area (MPa); B_1 is the blade width (m); h_p is the average cutting depth (m); G_t is the weight of the mound in front of the bulldozing plate; V is the volume of the mound in front of the bulldozing plate; μ_1 is the friction coefficient between soil particles; μ_2 is the friction coefficient between the soil and blade; θ is the slope (°); k_s is the loose degree coefficient of the soil; k_m is the fullness degree coefficient of the soil; H is the blade height (m); α_0 is the natural slope angle of the soil (°); X is the length of the worn blade contacting the ground (m); k_y is the cutting resistance per unit area after the blade is pressed into the soil (MPa); and δ is the cutting angle of the blade (°).

The powertrain model of the HETB is shown in Fig. 3.5, including a driver, working condition, engine-generator, motor drive system, vehicle dynamics, and control in MATLAB/Simulink. Equations (3.13) and (3.14) are used in the "Dynamic" module; Equations (3.1)–(3.3) are used in the "Diesel Engine" module; Equations (3.4)–(3.5) are used in the "Generator Drive" module; and Equations (3.6)–(3.8) are used in the "Motor Drive 1" and "Motor Drive 2" module.

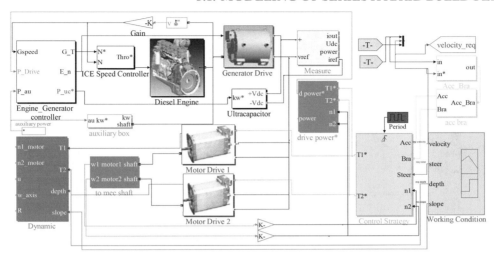

Figure 3.5: Powertrain model of the HETB.

3.1.5   MODEL VERIFICATION OF THE SERIES HETB

To verify the simulation model of the HETB, the values of all parameters and real working conditions (as shown in Fig. 3.6) were input into the simulation model to compare the simulation results with the real experimental data.

In Fig. 3.6, V (km/h) is the bulldozer velocity; depth (m) is soil-cut depth; slope (°) is road slope. The working process is: it is traveling during 1 s to 4 s; cutting soil during 4 s to 16 s; transporting soil from 16 s to 31 s; unloading the soil between 31 s to 33 s, and then traveling.

Figure 3.7 and Fig. 3.8 present the simulation and real velocity and drive forces of a single track under the operating condition in Fig. 3.6. Figure 3.9 indicates the simulation and real results of the engine output power. Figure 3.10 demonstrates the real efficiency of the generator and motor.

As it can be seen from Fig. 3.6 to Fig. 3.7, the velocities and driving forces from both the simulation and experimental data match fairly well, and the discrepancy of the single-track driving forces is about 1.45%.

In Fig. 3.8, the simulation and real engine output power agree well, except during the 4–16 s soil-cutting period. The simulation value is about 15–25 kw higher than the experimental value because the efficiency of the powertrain is different in simulation (hybrid) and experiment (traditional). The experiment transmission efficiency is approximately 0.9, which is higher than that of the simulation transmission. Figure 3.9 indicates the motor efficiency is below 0.90 at this period of the simulation. As a result, the engine must produce a higher power in the simulation than in the experiment in order to meet the same demand power.

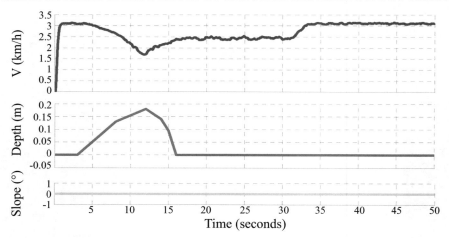

Figure 3.6: The real working condition of the bulldozer.

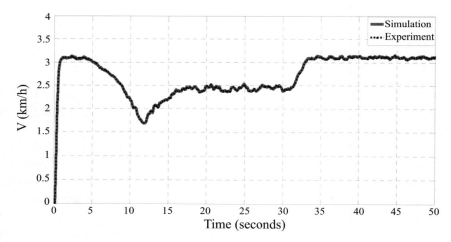

Figure 3.7: Bulldozer velocities in the simulation and experiment.

As mentioned previously, the simulation model is valid and can be utilized for further research, such as the control strategy development for the HETB.

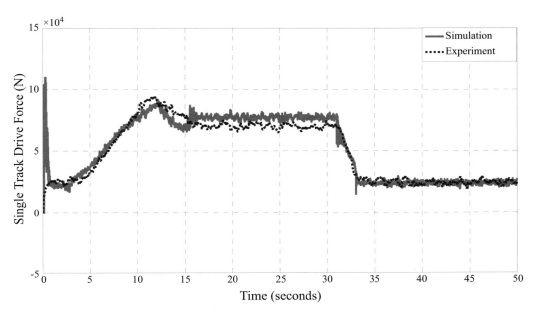

Figure 3.8: Simulation and experimental driving forces of a single track.

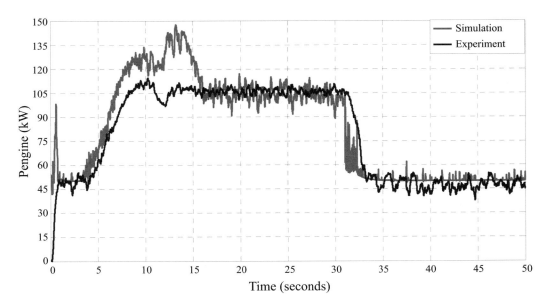

Figure 3.9: Simulation and experiment results of the engine output power.

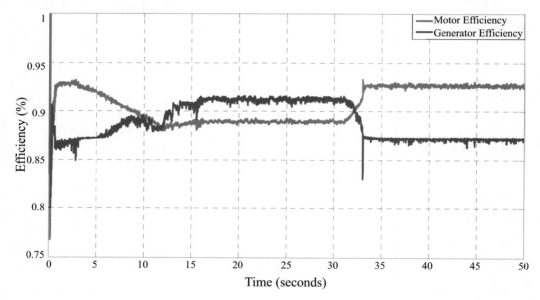

Figure 3.10: Generator and motor efficiency.

CHAPTER 4

Energy Management Systems for Electrified Heavy-duty Vehicles

In this chapter, we review energy management systems for hybrid vehicles and then design specifically such systems for heavy-duty vehicles.

4.1 REVIEW OF ENERGY MANAGEMENT SYSTEMS (EMSS) FOR HYBRID ELECTRIC VEHICLES

The existing EMSs of HEVs can be generally classified into heuristic strategies and optimization-based strategies. The most common EMSs employed in HEVs are shown in Fig. 4.1. In the following, these methods are reviewed in detail.

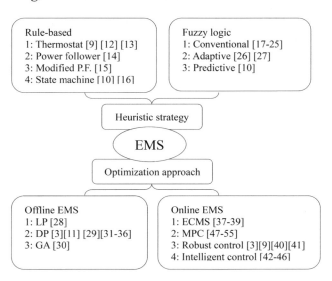

Figure 4.1: Classification of EMSs for HEVs.

4.1.1 HEURISTIC STRATEGIES

Heuristic strategies depend on a set of rules to determine the control action at each time instant. The rules are designed in accordance with intuition, human expertise, and/or mathematical models and, usually, without prior knowledge of any driving information. Deterministic rule-based and fuzzy logic approaches are two parts of this category. The latter has become considerably popular in recent years as artificial intelligence (AI)-related technologies develop [9, 11].

A. Deterministic rule-based strategies

Rule-based EMSs can be further classified into thermostat (on/off) strategy, power follower strategy, modified power follower strategy, and state machine-based strategy [10].

1. Thermostat strategy

The thermostat (on/off) control strategy is robust, simple, and easy to realize. In the thermostat strategy, the internal combustion engine (ICE) operates at its highest efficiency point once it turns on, while the battery's SOC is always maintained between its preset upper and lower bounds by turning on or turning off ICE. Due to the fixed rules, it lacks the ability to deal with uncertainties brought on by model inaccuracy and the flexibility in different drive cycles [9]. Accordingly, both drive cycle recognition [12] and prediction [13] are proposed to enhance the rule-based EMSs. Although the thermostat strategy provides the best efficiency for the engine-generator set, the overall system efficiency of HEV is low. Despite its simplicity, this strategy cannot satisfy the power demands of the vehicle at all operating conditions. Nevertheless, for a series hybrid electric city bus commuting on prescheduled routes, the thermostat control strategy is applicable.

2. Power follower strategy

The power follower control strategy is popular and has been successfully applied in commercial HEVs such as the Honda Insight and Toyota Prius [14]. Compared to the thermostat strategy, the power follower strategy is applicable to both parallel HEVs and series-parallel HEVs. However, the major disadvantage is that the overall efficiency of the powertrain is not optimal, and the emission control is not directly considered. The rules for the power follower control strategy are set up based on the following heuristics: (i) below a certain minimum vehicle speed, only the electric motor is used; (ii) if the demanded power is greater than the maximum engine power at its operating speed, the motor is used to produce the additional power; (iii) the motor charges the batteries by regenerative braking; (iv) the engine shuts off when the power demand falls below a limit at the operating speed to prevent inefficient operation of the engine; and (v) if the battery's SOC is lower than its minimum allowable value, the engine should provide additional power to replenish the battery via the electric motor/generator.

3. Modified power follower strategy

The modified power follower strategy, however, integrates energy usage and emissions into the cost function, which was proposed by Johnson et al. [15]. The main goal of this approach is to optimize both energy usage and emissions by the introduction of a cost function representing overall fuel consumption and emissions at all candidate operating points. This control strategy uses a time-averaged speed to find instantaneous energy use and emission targets.

4. State machine-based strategy

The state machine-based approach was proposed by Phillips et al. [16]. In this strategy, the transition between operating modes, such as ENGINE, BOOST, CHARGING, etc., is decided by a state machine that is based on vehicle operating conditions, change in driver demand, and any system fault [10]. Furthermore, it was claimed that the dynamic control algorithms generate output commands to each subsystem, e.g., desired torque output commands to each subsystem, desired torque from the engine. Implementation of a vehicle controller through state machines facilitates fault resilient supervisory control of the whole system. Nevertheless, optimization of the performance objectives such as fuel economy or emissions are not guaranteed. Furthermore, it is not clear how the proclaimed dynamic controllers are designed. Therefore, from an energy management point of view, this approach has no added value to conventional deterministic rule-based methods. Consequently, it seems that switching to fuzzy rule-based methods is a wise decision.

B. Fuzzy logic approach

The fuzzy logic theory is unique in its ability to simultaneously handle numerical data and linguistic knowledge. Fuzzy sets represent linguistic labels or term sets such as slow, fast, low, medium, high, and so forth. Fuzzy logic is a form of multivalued logic derived from fuzzy set theory to deal with reasoning that is approximate rather than precise. Fuzzy logic enables the development of rule-based behavior. The knowledge of an expert can be coded in the form of a rule and used in decision making. Looking into a hybrid powertrain as a multi-domain, nonlinear, and time-varying plant, fuzzy logic seems to be the most logical approach to the problem. In fact, instead of using deterministic rules, the decision-making property of fuzzy logic can be adopted to realize a real-time and suboptimal power split. In other words, the fuzzy logic controller is an extension of the conventional rule-based controller. The main advantages of the fuzzy logic approach are its robustness to measurement noises and component variability along with its adaptation [17, 18, 19]. In essence, a fuzzy logic controller is a natural extension of many rule-based controllers implemented in numerous vehicles today. Fuzzy logic-based methods are intensive to model uncertainties and are robust against the measurement of noise and disturbances but require a faster microcontroller with larger memory. As a result, the fuzzy logical approach is very suitable for multi-domain, nonlinear, time-varying systems such as HEVs [20, 21]. This strategy can be further categorized into conventional, adaptive, and predictive strategies.

1. Conventional fuzzy control strategy

   One of the conventional fuzzy approaches was to utilize a sophisticated controller composed of two fuzzy logic controllers such as a driver's intention predictor and power balance controller [10, 22]. Another one was designed by load-leveling such as in [23, 24, 25].

2. Adaptive fuzzy control strategy

   Adaptive neural fuzzy inference system [26] and machine learning algorithms [27] were introduced to the fuzzy logic technique to improve its robustness to variations of drive cycles. This strategy can simultaneously optimize both fuel efficiency and emissions. However, fuel economy and emissions are conflicting objectives, which means that an optimal solution cannot be achieved by satisfying all the objectives. The optimal operating point can be obtained using a weighted-sum approach to optimization of the conflicting objectives. Due to various driving conditions, appropriate weights have to be tuned for fuel economy and emissions. Considering stringent air pollution laws, operating points with high emissions are heavily penalized. The conflict within the adaptive fuzzy logic controller includes fuel economy, NO_x, CO, and HC emissions. In order to measure the relationship of the four contending optimizing objectives with a uniform standard, it is essential to normalize the values of fuel economy and emissions by utilizing the optimal values of fuel consumption and emissions at the current speed. This control strategy is able to control any one of the objectives by changing the values of the relative weights. Furthermore, tremendous reduction in vehicle emissions is achieved with negligible compromise in fuel economy.

3. Predictive fuzzy control strategy

   The fuzzy predictive strategy was developed at Ohio State University [10], and this strategy decides the future state according to the historical data through a look-ahead window along a planned route or speed predictions from traffic conditions along the route. If the information on the driving trip is already known, it is trivial to obtain a global optimum solution to minimize fuel consumption and emissions. However, the primary obstacles entail acquiring further information on planned driving routes and performing real-time control. This problem can be resolved using a global positioning system (GPS) to easily identify the probable obstacles like heavy traffic or a steep grade. The control strategies can be developed for specific situations; for example, if a vehicle is running on a highway and will enter a city (where heavy traffic may be encountered), it is advised to restore more energy for future use by charging the batteries. General inputs to the predictive Fuzzy logic control are vehicle speed variations, the speed state of the vehicle in a look ahead window, and elevation of the sampled points along a predetermined route. Based on the available history of vehicle motion and its variability in the near future, fuzzy logic control determines the optimal torque that the ICE contributes to the current vehicle speed. The predictive fuzzy logic control outputs a normalized GPS signal in $(-1, +1)$, which informs the master controller to charge or discharge the batteries and to restore enough energy for future vehicle operating modes.

4.1.2   OPTIMIZATION APPROACH

Optimization approaches rely on analytical or numerical optimization algorithms which are, obviously, able to optimize the performance [10]. The optimization method can be sorted into two main groups: offline optimization and online optimization.

A. Offline optimization

Based on the knowledge of past or future power demands, the offline optimization-based approaches aim to minimize the cost function, which reflects the fuel economy and/or emissions over a fixed and known drive cycle. Usually, they are useful and beneficial for design or comparison purposes as the benchmarks, and examples of such methods include linear programming (LP), dynamic programming (DP) [29], genetic algorithm (GA) [30], and simulated annealing (SA).

1. Linear programming

Fuel economy optimization is a convex nonlinear optimization problem, which is approximated by the linear programming method. Linear programming is mostly used for fuel efficiency optimization in series HEVs. Formulation of the fuel efficiency optimization problem using linear programming may result in a global optimal solution. An optimized design and control of a series hybrid vehicle by controlling the gear ratio and torque are proposed in [28]. The problem is formulated as a nonlinear convex optimization problem and approximated as a linear programming problem to find the fuel efficiency.

2. Dynamic programming

Dynamic programming (DP) usually depends on a model to provide a provably optimal control strategy by searching all state and control grids exhaustively [11, 31, 32]. However, DP is not applicable for real-time problems since the exact future driving information is seldom known in real applications [33]. Nonetheless, the DP-based strategy can provide a good benchmark for evaluating the optimality of other algorithms and contribute to improving real-time strategies [34, 35]. A DP-based approach to reduce the fuel consumption of a parallel hybrid electric truck is reported in [36].

Contrary to the rule-based algorithm, the dynamic programming approach usually relies on a model to compute the best control strategy. Dynamic programming guarantees global optimality through an exhaustive search of all control and state grids. It breaks the optimization problem into a sequence of decision steps over time. The optimization target is to minimize or maximize the objective function $J(x, u)$. The state of the system can be discretized into the state grid. At time step t_k, the system state x_k can be driven by control input u_k into another state on the next time step t_{k+1}. The one-step cost from t_k to t_{k+1} is defined as $J_{k \to k+1}$ and the accumulated cost from time step t_k to t_N is defined as $J_k = J_{k \to k+1} + J_{k+1}^*$. J_{k+1}^* is the optimal accumulated cost from t_{k+1} to t_N. The objective of the DP algorithm is to find the best control inputs u_k^* that minimizes J_k at every time step k so that the trajectory of the state from every initial point will be guaranteed as optimal. This procedure is performed through an iterative backward optimization. The resulting u_k^* is saved as a function dependent on x_k.

3. Genetic Algorithm

The Genetic Algorithm (GA) is a heuristic search algorithm to generate the solution to search problems and optimization. GA begins with a set of solutions called a population. The solutions from one population are taken according to their ability to form new populations. The most suitable solutions will have a better chance to grow over the poorer solutions, and the process is repeated until the desired condition is satisfied. GA is a robust and feasible approach with a wide range of search space to rapidly optimize parameters using simple operations. They are proven to be effective in solving complex engineering optimization problems characterized by nonlinear, multimodal, nonconvex objective functions. GA is efficient at searching the global optima without getting stuck in a local optimum. Unlike the conventional gradient based method, the GA technique does not require any strong assumptions or additional information about objective parameters. GA can also explore the solution space very efficiently. However, this method is very time consuming and does not provide a broader view to the designer.

A genetic algorithm is a powerful optimization tool that is particularly appropriate to multi-objective optimization. The ability to sample trade-off surfaces in a global, efficient, and directed way is very important for the extra knowledge it provides. In the case where there are two or more equivalent optima, the GA is known to drift toward one of them in a long-term perspective. This phenomenon of genetic drift has been well observed in nature and is due to populations being finite. It becomes more and more important as the populations get smaller. A non-dominated sorting genetic algorithm (NSGA) varies from GA only in the way that the selection operator works. Crossover and mutation operations remain the same. This is similar to the simple GA except in the classification of non-dominated fronts and sharing operations. A multi-objective genetic algorithm (MOGA) is a modification of GA at the selection level.

MOGA may not be able to find multiple solutions in cases where different Pareto-optimal points correspond to the same objective.

Reference [37] uses MOGA to solve an optimization problem for series HEV. The control strategy based on MOGA is flexible, multi-objective, and gives a global optimal. A MOGA is further used by [38, 39] to solve the optimization problem of HEVs where it optimizes the control system and powertrain parameters simultaneously to yield a Pareto-optimal solution. Montazeri-Gh et al. [40] present a genetic-fuzzy approach and find an optimal region for the engine to work. It provides an optimal solution to the optimization problem. Wimalendra et al. [41] applied GA to a parallel HEV to find the optimal power split for improved vehicle performance, and it showed promise in providing the maximum fuel economy for the known DC. Reference [42] implemented a fuzzy control strategy for reduced fuel consumption and emissions optimized by GA. MOGA is developed to reduce fuel consumption and emissions while optimizing powertrain component sizing [43]. Using NSGA, a Pareto-optimal solution is obtained for reduced component sizing, fuel consumption, and emissions [44].

B. Online optimization

Contrary to the offline optimization, the online optimization consists of an equivalent consumption minimization strategy (ECMS), MPC, robust control approach, and intelligent controls, etc.

1. Equivalent Consumption Minimization Strategy

The concept of ECMS was proposed by Paganelli et al. in [45] as a way to convert the global minimization problem to an instantaneous minimization problem solved at each step. The idea behind this is that the total fuel consumption is calculated as the sum of the real fuel consumption by ICE and the equivalent fuel consumption of the electric motor. This allows for a unified representation of both the energy used in the battery and the ICE fuel consumption. Using this approach, the equivalent fuel consumption is calculated in real time as a function of the current system-measured parameters. No future predictions are necessary and only a few control parameters are required. These parameters may vary from one HEV topology to another as a function of the driving conditions. ECMS can compensate for the effect of uncertainties of dynamic programming. The only disadvantage of this strategy is that it does not guarantee the charge sustainability of the plant.

The equivalent fuel consumption is calculated based on the assumption that any SOC variation in the future is compensated by the engine running at the current operating point. Paganelli et al. implemented an ECMS for a hybrid-electric sport utility vehicle in charge sustaining mode to minimize the fuel consumption and pollutant emission [46]. In [47], ECMS was used to minimize the fuel consumption of an

HEV by splitting the power between the ICE and the electric motor. It achieves a reduction in fuel consumption by 17.5% when compared to a solely ICE-based vehicle. Cui et al. developed an energy management strategy that comprises of two stages in [48]: (1) instantaneous optimization using ECMS and (2) global parameter estimation using DP. Knowledge of the distance of the next charging station during travel gives noteworthy fuel economy, and full knowledge of terrain preview gives almost 1% in fuel economy improvement. Tulpule et al. [49] propose an ECMS, which requires knowledge of the total trip distance instead of driving pattern information to improve fuel economy. References [50, 51] implement a modified ECMS for a series hybrid vehicle configuration with two different energy sources, and this is a generalization of the instantaneous ECMS proposed in [52, 53]. Supina and Awad [54] suggest turning on/off the engine according to the battery energy level, which results in an improved fuel efficiency of 1.6% to 5% over the thermostat control. [55] presents a parallel HEV without known future driving conditions for the development of a real-time control of fuel consumption. It uses ECMS 10 International Journal of Vehicular Technology for the instantaneous optimization of the cost function and it depends only upon the current system operation. Won et al. [56] propose an energy management strategy for the torque distribution and charge sustenance of the HEV using ECMS. In this, a multi-objective torque distribution strategy is first formulated and then converted into a single objective linear optimization problem.

2. Robust Control

   The robust approach is a type of output feedback control whose parameters are tuned in such a way that the matrix norms of closed loop systems are within the desired boundaries. Pisu et al. developed this strategy to solve the energy management problem in plug-in HEVs (PHEVs). In this approach, the objective was to determine an output feedback controller that minimizes fuel consumption with respect to a family of possible torque/power input profiles [57, 58]. However, robust control can only obtain a sub-optimal solution like other real-time optimization energy management strategies. Furthermore, robust control requires much effort in the manipulation of system equations. Mathematical complexity as well as simplification of a nonlinear time-varying system to a linear time-invariant system has prevented further development of robust control in the field of energy management for HEVs.

3. Intelligent control strategy

   Intelligent methods take reasonable decisions through a simulation of the human brain based on qualitative and quantitative information of the controlled system, which is well fitted to the control of complex nonlinear systems. Among the intelligent strategies, machine learning algorithms are most widely applied for the energy management control of HEVs, and these algorithms include NN [59], Elman neural network (ENN) [60], support vector machine SVM [61], and recursive least square

(RLS) [62] among others [63]. For machine learning approaches, accurate powertrain models are no longer required so the computational effort is significantly reduced.

4. Model predictive control

In recent years, as a powerful receding horizon control technique, conventional/adaptive/robust MPC has been widely adopted in the automotive industry (including HEVs) because of its capability to achieve optimal control in the presence of multivariable states/inputs constraints and good applicability in real vehicles. There is a trade off between DP and ECMS when HEVs adopt MPC [64]–[68], which solves the optimization problem of the energy management at every sampling instant by quadratic programming, Pontryagin's minimum principle, nonlinear programming, or stochastic DP [69, 70, 71, 72].

As shown in Fig. 4.2, MPC is generally implemented in the following three steps: (1) obtain the optimal trajectory according to the dynamic model of the system in the predicted horizon to minimize a predefined objective function in the presence of multivariable constraints; (2) apply the first control element of the optimal controller sequence with the minimum cost to the controlled system; and (3) move the whole prediction horizon window one sampling step forward and repeat the previous steps again [66].

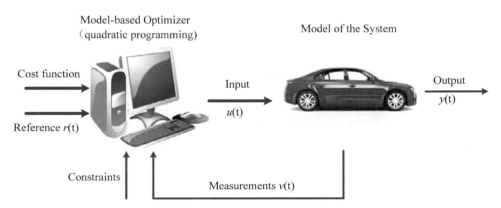

Figure 4.2: MPC formulation flow diagram.

MPC is an optimal control algorithm, originating from the chemistry industry. At the beginning it could only deal with control systems with slow dynamics, which tolerate time-consuming optimization calculation. For one certain HEV, the general dynamics model can be presented by

$$\left\{ \begin{array}{l} x(k+1) = A(k)x(k) + B(k)u(k) + n(k) \\ y(k) = C(k)x(k) + D(k)u(k) + v(k) \end{array} \right\} \tag{4.1}$$

where $x(k)$ denotes states, $u(k)$ represents control inputs, $y(k)$ refers to the outputs, $n(k)$ means state noise, and $v(k)$ indicates the measurement noise. An objective function can be adopted to assess the control performance combining the tracking error and control input. The objective function to be minimized can be expressed to achieve the optimal fuel economy or reduction of emissions as follows:

$$
\min_{u_0, u_1, \ldots u_{N-1}} J
$$

$$
= \min_{u_0, u_1, \ldots u_{N-1}} \sum_{i=1}^{N-1} \left[w_{i+1}{}^y \| y(k+i+1 \,|k) - y_{ref}(k+i+1\,|k) \|^2 \right.
$$

$$
\left. + w_i^u \| u(k+i\,|k) \|^2 \right]
$$

$$
= \min_{u_0, u_1, \ldots u_{N-1}} \sum_{i=1}^{N-1} \left[y(k+i+1\,|k) - y_{ref}(k+i+1\,|k) \right]^T Q \qquad (4.2)
$$

$$
\left[y(k+i+1\,|k) - y_{ref}(k+i+1\,|k) \right] + u(k+i\,|k)^T \, Ru(k+i\,|k)
$$

$s.t.$

$$
y_{\min} \le y(k) \le y_{\max}, k = 0, 1, \ldots N-1
$$
$$
u_{\min} \le u(k) \le u_{\max}, k = 0, 1, \ldots N-1
$$

where N represents the prediction horizon length, and w^y and w^u indicate the weights for the output y and control input u, respectively. An optimization algorithm is employed to calculate the optimal control input to minimize the above objective function considering the required constraint conditions. However, only the first control element in the obtained control sequence is used in the closed-loop system. After that, the prediction horizon moves forward by one sample, and the optimization process is implemented again with updated measurements.

From previous sections, it can be seen that the future driving information is a prerequisite for MPC formulation. Therefore, it is necessary to study the prediction methods used in the existing literature. A review on driving condition prediction was conducted in [73], where two categories were presented. One is the GPS and ITS-based prediction; whereas, the other one is a statistics and cluster analysis-based method. However, a more detailed survey will be carried out in this section by sorting and elaborating all the methods used for prediction of the future information in the existing literature. Accordingly, based on these methods, the MPC strategies are categorized in Fig. 4.3.

(a) Frozen-time MPC

The frozen-time MPC (FTMPC) [74, 75] adopts the same power demand in the whole prediction horizon, but it is usually used for comparison with other controllers

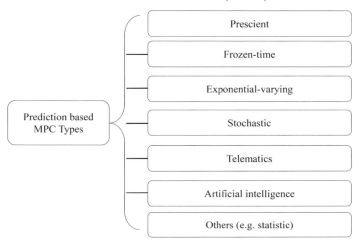

Figure 4.3: MPC types based on drive cycle prediction.

due to the poor control performance. Its performance is heavily dependent on the drive cycle and the prediction horizontal length. If the drive cycle changes sharply, its performance will be worse. Therefore, the method performs better on the highway than during urban driving. The other kinds of MPC developed later have better fuel economy compared with FTMPC. A series HEV was studied by using different types of MPC EMSs. The results demonstrate that the Frozen MPC consumes 29.8% more fuel than the prescient MPC (PMPC) under the examined drive cycle [74].

(b) Prescient MPC

Like the FTMPC, the PMPC is also used as a benchmark for comparison. PMPC perform similarly to the other developed MPCs which are also based on the assumption that all the future information can be obtained accurately. Even though PMPC solves the optimization problem based on the current cost and future cost-to-go obtained over a finite prediction horizon, the solution provided is suboptimal. In order to be a good basis, its parameters should be tuned and the performance should be verified by the global optimal one found by DP. Comparison results between the PMPC and DP in [75] showed that the MPC can achieve 96 % of the optimality of the DP. PMPC cannot be applied in practice due to its unrealistic assumption, but it is suitable to evaluate any MPC-based energy management strategy [76, 77, 78, 79, 80].

(c) Exponential varying MPC

The exponentially varying MPC adopts an exponential varying variable to reflect the future information in the prediction horizon. In the development process of this MPC, the unknown torque of the driver demand is assumed to be exponentially decreasing in the prediction horizon.

$$T_d (k + i) = T_d (k) e^{\left(\frac{-it_s}{t_d}\right)} \qquad i = 1, 2, \ldots N_p \qquad (4.3)$$

where $T_d(k)$ is the demand torque at the current step which is known; the demand torque $T_d(k + i)$ is in the prediction horizon; t_s and t_d are the sample time and decay rate, respectively. N_p denotes the prediction horizontal length. Based on the above equation, the velocity trajectory can also be calculated by the dynamic equation of the vehicle model. However, the vehicle velocity can be first determined by

$$V (k + i) = V (k) \times (1 + \varepsilon)^i \qquad i = 1, 2, \ldots N_p \qquad (4.4)$$

where $V(k)$ is the vehicle velocity, and ε represents the exponential coefficient. However, the accuracy of this prediction method relies on parameter values, which vary with the drive cycles. The effect of different values of ε on the prediction accuracy and fuel consumption of a power-split HEV was studied in [81]. However, this prediction method relies on the exponential-varying assumption of the future driving information, which impedes its extensive application if this assumption is violated.

(d) Stochastic MPC

The Markov chain, also called the stochastic, DP is usually utilized to deal with decision-making issues considering multi-period stochastic circumstances [82]. Since the Markov chain is a promising and important method utilized in modeling driver behavior or predicting the vehicle's velocity and power demands, much effort has been expended on developing optimization-based EMSs by using such a method.

Position-dependent and non-position-dependent discrete time Markov chains [83] were developed to predict vehicle states based on the information acquired by a GPS module with a traffic-flow information system. The Markov chains were integrated into the DP algorithm to form stochastic optimal energy management controllers for a parallel HEV. The results were compared to the DP to demonstrate their optimality and applicability. A stochastic MPC (SMPC) was designed in [3] for a series HEV, where the driver's future power demand was modeled as a Markov chain. Its performance with a known power-demand was compared with that of a PMPC, and an FTMPC using a constant power demand in the forecast horizon. The authors showed that the proposed MPC provided a fuel economy similar to the PMPC. A discrete-time Markov chain [84] was proposed [85] to predict the distribution of future power demands based on the torque demand and vehicle speed at the present step [86] by using the maximum likelihood estimation approach. The driver behavior was modeled as a Markov chain based on the several standard cycles [67]. Then, a shortest path stochastic DP was designed and implemented in a test vehicle to evaluate its performance. The results showed that the fuel economy was improved with fewer engine on/off events. As an improvement in the process of real-time implementation, an SMPC was proposed in [74]. This SMPC can handle a larger state dimension than the stochastic DP. A stationary Markov chain was utilized to generate or predict drive torque demands in [87]. A methodology from the techniques of cluster, evaluation, and Markov chain was designed for a plug-in hybrid electric bus on a fixed route to predict the future driving information. In this method, three clusters of the drive cycle segments were obtained according to some preselected feature variables, and then, a hidden Markov model was proposed to reconstruct the drive cycle for predictions [88]. A Markov chain that represents the power requested was taught to enhance the prediction capabilities of the MPC. Due to its ability to learn the pattern of the driver behavior, the method shows a performance similar to the PMPC. An SMPC was proposed in [89] for a parallel HEV, particularly, one running in a hilly region with traffic lights. This MPC considers the road grade and maintains the SOC inside its boundary to avoid degrading the energy efficiency. A finite-horizon Markov decision process was modeled and integrated into the MPC, which was compared to an ECMS and DP to verify its fuel economy. In [90], a driving-behavior-aware SMPC was developed for hybrid electric buses, where the K-means was used to categorize driving behaviors, and the Markov chains were utilized to model the driver behavior. A multi-step Markov prediction method was chosen for the vehicle velocity prediction based on the basis of the assumption that the vehicle states in the future are only dependent on the current ones instead of the previous or historical information, and then, the velocity predictor was used for MPC energy management controller development for a plug-in HEV [91]. A Markov chain was used and the transition possibility was calculated based on a dataset, including two real collected driving cycles and six standard driving cycles. The prediction precision was analyzed and compared to other methods such as the exponentially varying methods and neural networks. However, for stochastic control based on Markov chain, the control law is optimal only for the specific Markov chain. That is, the transition probabilities in the Markov chain are affected by the collected driving cycles [92]. If there is difference between the real driving condition and the collected data, the accuracy of the prediction will be degraded, and the optimality cannot be guaranteed by the algorithm.

Besides the Markov chain method, other stochastic approaches have also appeared in the literature. The future driving conditions were classified as external-sensor-based and stochastic methods in [93], where a stochastic approach was proposed by clustering and fitting the power demands of a fixed drive cycle to two normal distributions. An assumption that the vehicle will be operated according to these known distributions was used. The sequence of required power can be represented by the probability density according to the normal distribution. Therefore, the current driving condition can be detected by comparing the probability density. This method is adaptive since it updates the probability functions online. A stochastic method was presented in [94] for the EMS of fixed-route HEVs, where the fixed-route historical data was used, and a stochastic optimal power consumption model for HEV based on the road segment was developed by considering the SOC and fuel consumption as random variables on each road segment. In [95], an improved neighbor regressor was adopted to produce samples of the drive cycle, and MPC was used based on the obtained information. A load predictor was proposed in [96], where the K-nearest neighbor approach matches the present vehicle state to the historical training data in order to get a weighted set of predictions.

(e) Artificial intelligence MPC

Due to their strong ability in modeling, learning, and predicting [97], much effort has been put toward forecasting drive cycles by using artificial intelligence, such as neural network, Bayesian algorithm, fuzzy logic pattern recognition, decision tree, and support vector machine [98].

S. Jeon et al. in [99] employed neural network to identify the current vehicle driving patterns with 24 parameters, and determine the six predefined driving patterns which are closest to the current one. The superiority of the proposed technique was demonstrated by simulations. This approach was modified by reducing the feature parameters to 17 [100] and 15 [101], respectively. A drive pattern identifier [102] was presented by a learning vector quantization neural network to study six typical drive cycles. The parameters such as neuron number and sampling window were tuned by sample training simulations. Compared to the previous research, only seven parameters were utilized to represent drive patterns. Under an assumption that the vehicle's future load is available or derivable from the historical data, a recurrent neural network [103] was proposed to predict the temporary (next 20 seconds) vehicle load based on the previous load series. To recognize the driving pattern, a fuzzy logic pattern recognition (FL-PR) scheme [104, 105] was proposed to characterize the driving pattern. In [59] and [106], a machine learning approach based on the neural network was proposed to predict the traffic congestion, road type, and driving trends. At the same time, another neural network was designed to model the driver's response according to the predicted driving conditions. Simulation via Powertrain Systems Analysis Toolkit (PSAT) verifies the effectiveness of the trained neural network in predicting traffic congestion conditions, driving trends, and roadway type. There is no relevant relationship between the type of drive cycles and the driving block. A dual neural network was developed in [107] for drive cycle identification and prediction. The goal of the first one was to identify the practical drive cycle to one of the eleven standard drive cycles; whereas, the latter one was adopted to predict the driving trend in the future. A fuzzy logic controller (FLC) was used to divide the driving blocks into three categories (low, medium, and high speeds) instead of the whole cycles proposed in [108]. A neural network-based trip model [109] was designed to improve the gas-kinetic based trip model [110] for highway drives, which was trained and validated by the available data from WisTransPortal. After comparing three neural networks (radial basis function neural network, wavelet neural network, and recurrent neural network), the dynamic recurrent one was chosen and proposed to predict future drive cycles in [111]. A fuzzy logic controller [112] based on the T-S (Takagi-Sugeno) fuzzy theory was proposed to match the real-time drive cycle to one of the six typical drive cycles. In real drivings, it is impractical for a vehicle to accurately follow a predefined drive cycle in the presence of traffic conditions, weather, and preference of different drivers; many vehicles run on preset or fixed routes [94]. A drive cycle estimation algorithm was presented in [113] based on the neural network technique for service vehicles, such as refuse-collecting vehicles and delivery trucks, driving along the same routes but with different drive cycles. This method has been trained and verified by the real data and demonstrated with over 90% accuracy. A data-based approach was developed in [114] to estimate the future load demand, where a short-term load prediction was adopted with Bayesian inference, and a cycle detection method was designed [115] for real-time MPC application via correlation analysis. A radial basis neural network [116] was designed to predict the short-term future vehicle velocity through the historical velocity data. The forecast accuracy and length were also studied for the EMS implementation. An adaptive energy management was developed based on the reinforcement learning for a hybrid electric tracked vehicle in [117]. The integration of the AI-based drive cycle prediction method and MPC algorithm is shown in Fig. 4.4.

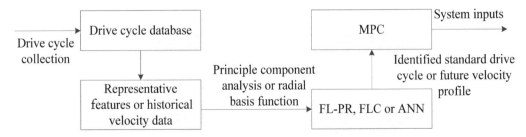

Figure 4.4: Integration of the AI-based MPC algorithm.

No matter the method used in the current literature, the AI-based drive cycle prediction can be divided into two subclasses: the first one defines characteristic parameters of the whole standard drive cycles or a combination of their parts (driving blocks); the current drive cycle can be matched to any one of the predefined standard drive cycles. The prediction accuracy of this approach heavily depends on the chosen typical or standard drive cycle used to train the algorithms. The second one is implemented with the assumption that the future drive cycle is related to the previous one or other historical data. Therefore, historical and current driving information is required to predict the temporary future information. Obviously, this method is dependent on how accurate this assumption is. In other words, if the drive cycle experiences dramatic changes, the prediction will become poor.

(f) Telematics

The aforementioned prediction problem can be alleviated to some extent by recent advances in intelligent transportation systems (ITSs) using onboard GPS, advanced traffic flow modeling techniques and geographical information systems (GISs) [118]. These vehicular telemetry technologies can be combined with other communication modules from vehicle-to-vehicle (V2V) or vehicle-to-infrastructure (V2I) to provide information on the upcoming traffic or road conditions for vehicles [118]. When such telematics-based technologies or systems are available, the information acquired by them can be exploited to improve prediction accuracy [120].

An MPC energy supervisory controller for HEVs was presented in [121]. A predictive SOC reference signal generator was developed by using the battery SOC and the measurement for an on-board navigation system. Only the average vehicle speed and road topographic profile are needed to generate the reference signal. The future speed profile is estimated in [122] based on GPS predictive information. A predictive controller for a power-split hybrid bus was proposed in [123]. The controller utilized data from GPS and historical data of the driving along a fixed route. It also mentioned that with access to the traffic information from GPS and GIS, accurate prediction of the propulsion load of the vehicle can be made. Since the route is predefined or fixed for buses or other service vehicles, the predictive control will be more attractive based on the acquired information from the onboard sensors. The benefits of incorporating traffic data into the energy management of plug-in HEVs were analyzed in [124]. A long-term SOC trajectory was planned by using time-varying traffic information to improve the MPC performance for power-split HEVs; in addition, a short-term velocity was predicted by a radial basis function neural network for MPC prediction in [125]. The measurements obtained from the GPS during vehicle operation were used to predict the speed profile which is relevant to the HEV's power demand over a fixed-route operation [126]. A parallel hybrid hydraulic vehicle with repeated routes was studied. A database of velocity trajectories and the corresponding vehicle position was updated after each drive on the specific route. According to the measurements of the GPS, the current location was utilized to match prediction profiles in the database. In the situations where multiple profiles were matched, the profile with the highest counter calculated in [65] was employed. The results showed that the proposed EMS with the prediction algorithm can obtain an extra 5% in fuel savings, which is quite close to that of the PMPC. The study that incorporates traffic information obtained from V2V and V2I into EMS development was conducted by [127]. A velocity and power trajectory was planned using the assistance of GPS, GIS, and the Global Navigation Satellite–based system (GNSS). The planned velocity trajectory can be an input of the active cruise controller; whereas, the power trajectory was used to develop the energy management system for hybrid electric trucks [128]. The velocity and power over an appropriate time horizon was predicted in [129] by using the information provided by GPS, V2V, V2I, and ITS systems. Then the predictions were used in the MPC energy EMS development. In addition, the sensitivity of the energy used by HEVs on the prediction errors was analyzed. A vehicle energy management system was presented by adopting the information of the upcoming topography and speed limitations to plan the speed and gear shifts for a truck [130]. The detailed online iteration procedure for trajectory prediction was elaborated upon. The Gipps' following car model [131] was adopted with the vehicle states (position and speed), the designed reference trajectory obtained by V2V, and the macroscopic traffic information (traffic speed and density). They depend on the traffic sensors or remote V2I traffic centers which use networked road site units (RSUs) to obtain traffic signal information. The real road and traffic information from V2V and V2I was used for optimal speed profile generation based on the energy optimization, and the online obtained optimal speed was recommended to the drivers for eco-driving [132]. The authors in [133] integrated the MPC energy management strategy into the connected HEVs, which simultaneously achieves the energy management optimization of the HEV powertrain, and the energy consumption minimization of the active cruise control by using the traffic data from the telematics. In other words, this work has optimized fuel efficiency and safety at the same time. A predictive energy management was proposed and applied in hybrid long-haul trucks. The strategy used information from the GPS and other information (e.g., the speed limits) to design battery charge/discharge profiles [134]. The proposed DP-based MPC strategy was evaluated under typical drive cycles. Dynamic traffic feedback data were used in the development of the MPC for a power-split HEV [135]. By using real-time traffic data, the battery SOC trajectory was generated and employed as the final-state constraint. The results showed that a better fuel economy could be achieved by using the traffic flow data. Different route prediction approaches were also discussed in [136].

Most of the above methods are emphasized in the predictive energy management development for HEVs running on a fixed or preset route. In this situation, the historical data can be collected by the telematics systems along with being processed and stored in a database for real-time matching with the current driving conditionings. With the fast development of ITS, autonomous vehicles and onboard telematics techniques, the vehicle reference trajectory can be obtained in advance, which contributes to the utilization of real-time MPC EMSs [137, 138]. A review was presented in [139] to summarize the vehicle energy management and saving approaches with the traffic data and information obtained by vehicle telematics. Focus was placed on three main aspects: (1) traffic supervision systems; (2) intelligent energy management systems of HEVs; and (3) intelligent charging management systems.

(g) Average-based MPC

MPC is an optimization-based control strategy, and for vehicle applications, it usually requires the information of the drivers' command (or drive cycle) in advance. However, in [147] Wang et al. proposed an average-based MPC that does not need any future driving information. The upcoming driving information is estimated by a moving average of the just past driving information. Take the urban dynamometer driving schedule (UDDS) for example. Figure 4.5 shows the vehicle velocity, acceleration, and power. For the sign of power, the positive sign means vehicle driving power, and the negative sign means the regenerative power.

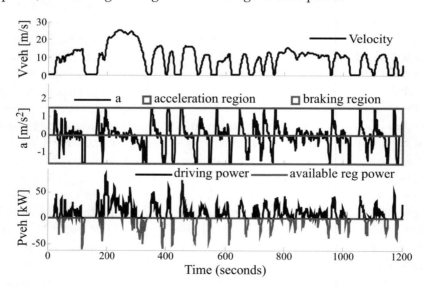

Figure 4.5: Driving information of UDDS.

From the figure of the vehicle power, we find the positive and negative powers roughly appear in pairs, and their amplitudes are basically identical. Therefore, it is reasonable to use a moving window with a suitable width to estimate the average power in the current region.

In Fig. 4.6, the window with a solid line represents the current moving window, and the window with a dotted line represents the window in the last step, where w and δt, respectively, means the width and interval of the moving window. The average vehicle power in the covered horizon is assigned to the current step of the moving window.

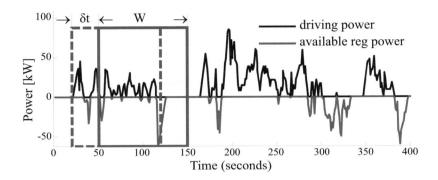

Figure 4.6: Average vehicle power estimation using the moving window.

(h) Others

Besides the aforementioned methods, other algorithms also appeared in the current literature. [140] developed an MPC that did not require the time-ordered prediction of the driving profiles but, instead, required a prediction of their distribution. The approach proposed was based on Pontryagin's minimum principle [141] under an assumption that the optimal costate is constant. Authors in [142] selected several typical drive patterns and some characteristic parameters of the power demand, which were used to classify each drive pattern. A set of deterministic rules with the thresholds of the characteristic parameters can be tuned by the selected drive patterns, and then, the real-time drive pattern can be matched to one of the selected drive patterns by using the rules and the current values of the characteristic parameters. Since a drive cycle can be represented by the frequency spectrum and probability density function, a method that used these representations was proposed in [143] to create alternative drive cycles, which are composed of similar information to the original drive cycle in terms of speed distribution and frequency spectrum. An optimal battery SOC trajectory was determined offline by using the navigation system and making it act as the terminal state reference for MPC application, where a simple statistic drive model was developed for the velocity prediction [144]. The statistic parameters such as average acceleration and standard deviation of the acceleration were utilized to predict the drive cycle in [145]. The Gaussian mixture model was also employed to identify the drive cycle by using the force applied on the gas and brake pedal [146]. Three different techniques for vehicle speed prediction, including the exponentially varying approach, stochastic approach, and NN-based approach were discussed in [81], where the parameter sensitivity was analyzed. Furthermore, the prediction accuracy, computational burden, and fuel economy were also compared.

4.2   ENERGY MANAGEMENT FOR ELECTRIFIED HEAVY-DUTY CONSTRUCTION VEHICLE

Hybrid electric construction machineries are different from road electric hybrid vehicles, mostly in three aspects: operation, key components, and reliability. (1) Operation: One representative working period of a 24-ton class conventional bulldozer is illustrated in Fig. 4.7. It can be seen that the demand power changes periodically and very quickly under typical conditions. The variation of its total load power can be from 30–180 kW. (2) Key components: Because of large fluctuations in power demands in construction vehicles (see Fig. 4.7) compared to road hybrid vehicles, an ultracapacitor as a supplementary energy source seems more suitable than a battery pack since a ultracapacitor can offer a quick power change to meet demands without seriously shortening its life. In the hybrid construction machinery, the engine, electric motor, and generator need to be installed together in a limited space, thus the electric motor is required to have high power density. (3) Reliability: Construction machineries are used in all types of environmental conditions, therefore its components need to be more reliable and robust than road vehicles.

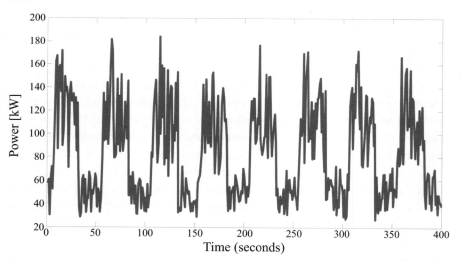

Figure 4.7: Load power in one typical working condition of a 24-ton class bulldozer.

According to the differences between hybrid electric construction vehicles and HEVs, the control strategy applications should also be different. An energy management strategy using the fuzzy logic algorithm is proposed for a new hybrid bulldozer in [148], which is verified through HIL (hardware in loop) simulations. In [149], a control strategy based on the dynamic work point was developed to improve the system performance of hydraulic excavators, which deals with the working point of the engine dynamically, and the advantages and disadvantages of two different control approaches were researched. In [150] Lin Xiao et al. developed a dynamic control strategy considering the multiple work points for the parallel hydraulic hybrid excavator.

4.2.1   RULE-BASED ENERGY MANAGEMENT

A power control technique is proposed to regulate the power supply and combine the work features of the hybrid electric tracked bulldozers considering the relationship among the engine generator, motors, and ultracapacitor. The engine output power is controlled to follow the load power demands of the bulldozer; however, when there is a power shortage caused by the excessive demand of the load power, the ultracapacitor will handle that as an auxiliary power source. The working points of the engine-generator are calculated according the load power demands and ultracapacitor SOC.

The allocation strategy of the power is shown in Table 4.1, where P^* means the mechanical power of the target demand, P_{dc} means the electric power of the DC bus demand, P_g means the output power of the generator, P_{uc} means the ultracapacitor power; P_e, P_{e-max}, P_{e-min} represents the output power, maximum output power, and minimum output power of the engine, respectively; η_1 represents the generator power efficiency; SOC_{max} and SOC_{min} are the maximum and minimum state of charge (SOC) of the ultracapacitor, respectively. P^* can be obtained from Section 4.3 as:

$$P^* = \frac{P^*_{Track}}{\eta_{E-T}} = \frac{F^*_{Track}r}{\eta_{E-T}} = \frac{(F_E + F_T)^*r}{\eta_{E-T}} \tag{4.5}$$

Where P^*_{Track} is the track demand power; η_{T-E} is the engine-to-track transmission efficiency; F^*_{Track} is the motion resistance.

Table 4.1: Power allocation strategy

Judgment	Ultracapacitor State	Power Supply
$P^* < P_{e\_max}$ $SOC < SOC_{max}$	Charging	$P_g = \eta_1 * P_e$ $P_{uc} = P_{dc} - P_g$
$P^* < P_{e\_max}$ $SOC \geq SOC_{max}$	Not-working	$P_g = \eta_1 * P_e$ $P_{uc} = 0$
$P^* > P_{e\_max}$ $SOC > SOC_{min}$	Discharging	$P_g = \eta_1 * P_{e\_max}$ $P_{uc} = P_{dc} - P_g$
$P^* > P_{e\_max}$ $SOC \leq SOC_{min}$	Not-working	$P_g = \eta_1 * P_{e\_max}$ $P_{uc} = 0$

Figure 4.8 shows the block diagram of the power allocation strategy. The driver's intention is considered as the target demand mechanical power P^*, and the vehicle management system (VMS) generates the required power of the drive motor P_m^* according to the vehicle working conditions. At the same time, the VMS provides the target speed of the engine n^* and the target current of the ultracapacitor I^*. The engine-generator yields the output power P_{eg}, and the ultracapacitor yields the output power P_{uc} based on a lookup table. The output power of the engine-generator and ultracapacitor is then transmitted to the motors through DC Bus.

In this control strategy, the parameters shown in Table 4.2 need to be properly optimized to minimize the fuel consumption to satisfy the operation demand.

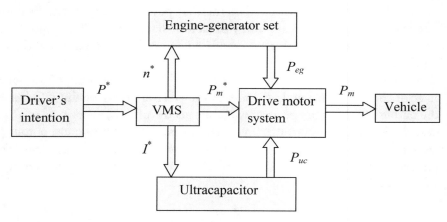

Figure 4.8: Power allocation flowchart.

Table 4.2: Control strategy parameters

Parameters	Instructions
SOC_{max}	Maximum SOC
SOC_{min}	Minimum SOC
N_1	Working speed 1 of the engine
N_2	Working speed 2 of the engine
N_3	Working speed 3 of the engine

Engine Control Strategy

A multi-point switching control approach is adopted here to regulate the engine speed, which is calculated by the demand power of the bulldozer load. The engine operates at high speeds when the load demand power is high, and operates at low speeds when the load demand power reduces.

Figure 4.9 shows the scheme of the switching control strategy for the engine speed, where the x-axis and y-axis represent the engine speed and load demand power, respectively. The y-axis is divided into three parts: low, medium, and high load demand powers, where the engine speeds in these three areas are denoted as N_1, N_2, and N_3 in sequence. The power hysteresis band is added between the adjacent areas to avoid the frequent switching of the engine speed.

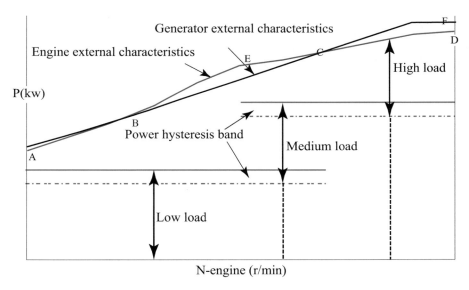

Figure 4.9: Scheme of the switching control strategy for the engine speed.

Experimental Verification of the Proposed Control Strategy
A test bench experiment was developed to validate the proposed power control strategy for the hybrid electric tracked bulldozer. The test bench consisted of an engine, drive motor system, generator, auxiliary power source, dynamometer, and the VMS, as shown in Fig. 4.10a. This test bench adopts the proposed power control strategy which was embedded in the VMS to regulate the power supply between the motors and engine-generator. The control interface of the dynamometer is shown in Fig. 4.10b, which manages the speed control of the dynamometer and motors. The load torque of the drive motor is controlled by the drive motor controller. In the control process, there are step changes for the drive motor load torque, as shown in Fig. 4.11, which is increased from 30–100 Nm, and then reduced to 30 Nm, in order to simulate the response of the bulldozer under sudden load changes. Note that the load torque is linear with respect to the accelerator pedal.

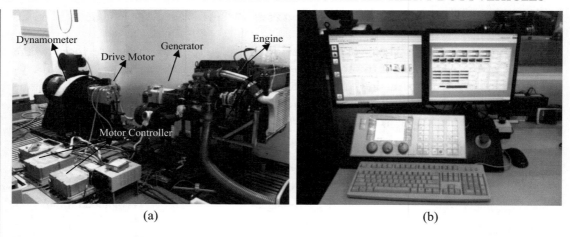

(a)   (b)

Figure 4.10: Hardware setup for experiments. (a) Structure of the hybrid bulldozer test bench; (b) Dynamometer control interface.

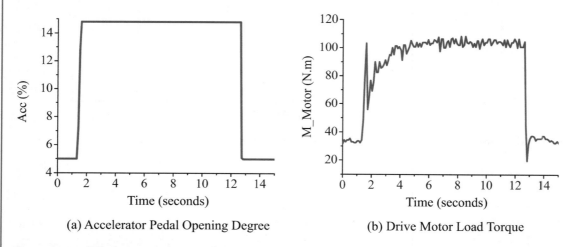

(a) Accelerator Pedal Opening Degree   (b) Drive Motor Load Torque

Figure 4.11: Working condition of the test bench.

From the above figure, we can see that the engine-generator output power tracks the load demand power well in the working conditions shown in Fig. 4.11. Thus, the proposed power tracking control strategy is effective in the optimal control for the hybrid electric tracked bulldozer.

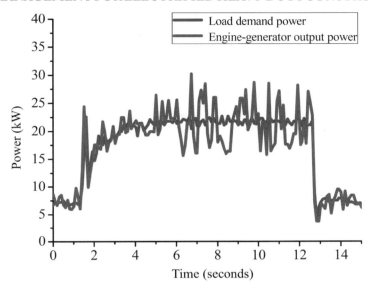

Figure 4.12: Load demand power and engine-generator output power.

4.2.2 DYNAMIC PROGRAMMING ENERGY MANAGEMENT

The DP-based control strategy needs to be conducted based on the discrete control variable and system plant, whose implement procedure is presented as follows.

Problem Formulation

The state/variables in the control loop need to be known to formulate DP. Here, the state is *SOE*, and the control input is the ultracapacitor output power. The discrete dynamics model of HEVs is

$$x(k + 1) = f(x(k), u(k)),$$ (4.6)

where $u(k)$ and $x(k)$ represent the control input and state, respectively. The sampling step is chosen as 1 second.

The objective of this optimization problem is to yield the optimal controller sequence $u(k)$ to minimize the fuel consumption over a receding moving horizon in the given driving cycle. A cost function for the optimization objective is defined as

$$J = \sum_{k=0}^{M-1} L(x(k), u(k))$$ (4.7)

where L means the instantaneous cost and M is the time length of the given drive cycle.

The physical constraints of the state and input are given by the following inequalities

$$\begin{cases} SOC_{\min} \leq SOC \leq SOC_{\max}; \\ SOE_{\min} \leq SOE \leq SOE_{\max}; \\ N_{e\_\min} \leq N_e \leq N_{e\_\max}; \\ P_{e\_\min} \leq P_e \leq P_{e\_\max}; \\ T_{e\_\min} \leq T_e \leq T_{e\_\max}; \end{cases} \qquad (4.8)$$

which are employed to guarantee the safe and smooth operations of the key components (i.e., the engine, ultracapacitor, and motor), and enable HEV to satisfy the speed/load requirements all the time.

Dynamic Programming Implementation

The main advantage of the DP is that it can deal with nonlinear systems subject to state/input constraints, which is described in the following steps:

Step $M - 1$:

$$J^*_{M-1}(x(M-1)) = \min_{u(M-1)} [L(x(M-1), u(M-1))] \qquad (4.9)$$

Step k, for $0 \leq k < M - 1$:

$$J^*_k(x(k)) = \min_{u(k)} \left[L(x(k), u(k)) + J^*_{k+1}(x(k+1)) \right] \qquad (4.10)$$

where $J^*_k(x(k))$ represents the accumulated optimal cost value from the time step t_k to the terminal time instant, $x(k + 1)$ refers to the state at the $(k + 1)$th time step when the control law u_k is applied at the time step t_k. The optimal control law is solved according to the above recursive equation. The minimization process is implemented subject to the equality constraints in the drive cycles and the inequality constraints shown in (4.8) simultaneously.

Procedure of Dynamic Programming

The DP procedure can be described with an example shown in Fig. 4.13, which refers to a single degree-of-freedom generic configuration of the HEV. UC SOE is chosen as the decision variable, which can be a limited number of values (i.e., 0.6, 0.65, or 0.7), given that the constraints on the state of energy can be satisfied easily. That is because the values of the *SOE* will not be considered except admissible ones, and the initial and final values are considered with no effort. The DP algorithm aims to calculate the optimal control sequence of the *SOE* to minimize the total cost. As the change of *SOE* is proportional to the integration value of the battery power between the sampling steps, determining a sequence for *SOE* and UC power is equivalent. The battery power constraints are formulated in terms of the maximal and minimal variations of the *SOE* between two adjacent time steps.

The first step for adopting the algorithm is to compute all the arc costs, which are the moving costs from time k to time $k + 1$ of all admissible nodes as shown in Fig. 4.13. Once all the arc costs have been fixed, the cost-to-go can be obtained, by going backward from the final node (Fig. 4.13b). For example, at time $k = N - 1 = 4$, all *SOE values* are admissible (H, I, K), but only the final node L is accepted, thus, three arc costs should be defined: $H \rightarrow L, I \rightarrow L$, and $k \rightarrow L$. At time $k = 3$, there are nine possible compositions (from any of E, F, G to any of H, I, K). Similar choices can be made for all other sampling steps. At time $k = 4$, the cost-to-go of each combination from H, I, K represents the arc cost since the optimization horizon ends at the subsequent time instant. At time $k = 3$, the cost-to-go of each node is the minimum cost from that node to the final node, thus the cost-to-go of the node E corresponds to the path with the minimum cost among the possible candidates: $E \rightarrow H \rightarrow L, E \rightarrow I \rightarrow L$, and $E \rightarrow K \rightarrow L$. From Fig. 4.13a, we know the respective costs are: $2 + 1.4 = 3.4, 2.3 + 1.9 = 4.2$, and $1.8 + 0.7 = 2.5$ which are shown in Fig. 4.13b to display the corresponding combination. In this sense, the cost of the best paths of $E \rightarrow K \rightarrow L, F \rightarrow K \rightarrow L$, and $G \rightarrow H \rightarrow L$ are 2.5, 1.6, and 1.4, respectively. Only this information is needed before the algorithm moves forward to the next time step ($k = 2$), and calculates the arc costs for the nodes B, C, and D. According to Bellman's optimality principle, the optimal path from E, F, or G to L will not be influenced by the combination of the preceding sampling step. Thus, the cost-to-go from B to L is equal to the sum of that from B to either E, F, G, and of the minimal cost from the chosen intermediate node to L. For example, $B \rightarrow E \rightarrow L$ costs 1.9 (cost of $B \rightarrow E$) plus 2.5 (minimal cost of $E \rightarrow L$). With the similar procedure, the arc costs of the entire graph of Fig. 4.13b can be obtained. After that, it can be concluded that $A \rightarrow B \rightarrow F \rightarrow K \rightarrow L$ is the optimal path, which has the lowest cost 4.9.

4.2.3   MODEL PREDICTIVE CONTROL (MPC)

The model predictive controller can be developed using the following equations according to the HEV model developed in the Section 4.3:

$$x_1 = -\frac{u(t)}{E_{cap}}$$
$$x_2 = B_e(P_e) = B_e \left(\frac{P_{uc} - P_{req}}{\eta_g} \right)$$

$$(4.11)$$

where $x_1 = SOE$ and $x_2 = B_e$ denote the fuel consumption; while, $u = P_{uc}$ represents the control input.

The vectors of control inputs, states, outputs, the measured inputs are defined as:

$$x = \begin{bmatrix} SOE \\ B_e \end{bmatrix}, \quad u = P_{uc}, \quad v = P_{req}, \quad y = \begin{bmatrix} SOE \\ B_e \end{bmatrix} \qquad (4.12)$$

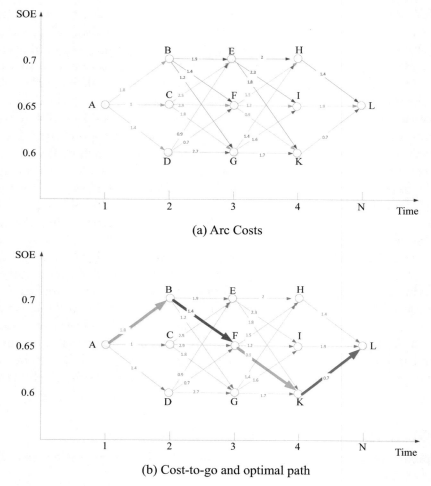

(a) Arc Costs

(b) Cost-to-go and optimal path

Figure 4.13: Single degree-of-freedom generic configuration of the HEV.

The discretized and linearized model of the powertrain is represented by:

$$\begin{cases} x(k+1) = A(k)x(k) + B_u(k)u(k) + B_v(k) \\ y(k) = C(k)x(k) \end{cases} \tag{4.13}$$

In this equation,

$$A = \begin{bmatrix} 1 & 0 \\ 0 & 1 \end{bmatrix}; \quad B_u(k) = \begin{bmatrix} -\frac{1}{E_{cap}} \\ -m_1 \end{bmatrix}; \quad B_v(k) = \begin{bmatrix} 0 \\ m_1 * P_{ref} + m_2 \end{bmatrix}; \quad C(k) = \begin{bmatrix} 1 & 0 \\ 0 & 1 \end{bmatrix}$$

The minization of the cost function can be described by:

$$\min_{u_0, u_1, \ldots u_{N-1}} J$$

$$= \min_{u_0, u_1, \ldots u_{N-1}} \sum_{i=1}^{N-1} \left[w_{i+1}{}^y \left\| y(k+i+1 \,|k) - y_{ref}(k+i+1 \,|k) \right\|^2 + w_i^u \left\| u(k+i \,|k) \right\|^2 \right]$$

$$= \min_{u_0, u_1, \ldots u_{N-1}} \sum_{i=1}^{N-1} \left[y(k+i+1 \,|k) - y_{ref}(k+i+1 \,|k) \right]^T Q \qquad (4.14)$$

$$\left[y(k+i+1 \,|k) - y_{ref}(k+i+1 \,|k) \right] + u(k+i \,|k)^T R u(k+i \,|k)$$

$s.t.$

$$y_{\min} \le y(k) \le y_{\max}, k = 0, 1, \ldots N - 1$$
$$u_{\min} \le u(k) \le u_{\max}, k = 0, 1, \ldots N - 1$$

where, N is the prediction horizon length; w^y and w^u refers to the weights for the output y and control input u.

The main objective of the optimal problem is to achieve optimal fuel economy while tracking the *SOE* reference value. The *SOE* reference trajectory is calculated from the DP optimization illustrated in previous section and zero is assigned to be the fuel consumption's reference trajectory. The state penalty Q and the input penalty R are:

$$Q = \begin{bmatrix} 1000000 & 0 \\ 0 & 1 \end{bmatrix}; \quad R = 10$$

To transfer the optimal problem into a quadratic form with regard to the control input, as the prediction horizon length is N, the trajectory of the future states will be calculated by the discrete model [148]:

$$\underbrace{\begin{bmatrix} x(k+1) \\ x(k+2) \\ \vdots \\ x(k+N) \end{bmatrix}}_{\bar{X}} = \underbrace{\begin{bmatrix} A \\ A^2 \\ \vdots \\ A^N \end{bmatrix}}_{S^x} x(k) + \underbrace{\begin{bmatrix} B_u & 0 & \cdots & 0 \\ AB_u & B_u & \cdots & 0 \\ \vdots & \vdots & \ddots & 0 \\ A^{N-1}B_u & A^{N-2}B_u & \cdots & B_u \end{bmatrix}}_{S^u} \underbrace{\begin{bmatrix} u(k) \\ u(k+1) \\ \vdots \\ u(k+N-1) \end{bmatrix}}_{\bar{U}}$$

$$+ \underbrace{\begin{bmatrix} B_v(k) \\ B_v(k) + B_v(k+1) \\ \vdots \\ B_v(k) + B_v(k+1) + \cdots B_v(k+N-1) \end{bmatrix}}_{V} \qquad (4.15)$$

$$\underbrace{\begin{bmatrix} y(k+1) \\ y(k+2) \\ \vdots \\ y(k+N) \end{bmatrix}}_{\bar{Y}} = \underbrace{\begin{bmatrix} C & 0 & 0 & 0 \\ 0 & C & 0 & 0 \\ 0 & 0 & \ddots & 0 \\ 0 & 0 & 0 & C \end{bmatrix}}_{C^x} \bar{X}$$

After we insert (4.15) into (4.14), we can get the convex quadratic objective function only with respect to the input:

$$J(x_0, u_0) = \frac{1}{2}\bar{U}^T H \bar{U} + F^T \bar{U}$$
$$H = 2(C^x S^u)^T \bar{Q}(C^x S^u) + \bar{R}$$
$$F = 2(C^x S^u)^T \bar{Q} \left(C^x S^u - \bar{Y}_{ref}\right) \qquad (4.16)$$

s.t.

$$\bar{U} \geq \max\left(\bar{U}_{\min}(U), \bar{U}_{\min}(\bar{U}), \bar{U}_{\min}(X)\right)$$
$$\bar{U} \leq \min\left(\bar{U}_{\max}(U), \bar{U}_{\max}(\bar{U}), \bar{U}_{\max}(X)\right)$$

In this equation, F is the gradient vector. H is Hessian matrix, which is positive and symmetric or semi-positive definite. $\bar{Q}, \bar{R}$, and $\bar{Y}_{ref}$ are related to the prediction horizon length N based on Q, R, and Y_{ref}.

qpOASES [164] as an open source solver is utilized to solve the energy management problem. $u_0, u_1, u_2 \ldots u_{N-1}$ as the optimal control input sequence is obtained from the solver qpOASES, then the first element of this trajectory u_0 is implemented to the plant model of the HETV. In the subsequent step, the value of the state will be updated. The optimal problem is solved by repeating this procedure during subsequent time steps.

4.2.4   CASE STUDY

In order to evaluate the three energy management strategies aforementioned in the previous section, two scenarios are utilized and compared in this section: typical working condition and working condition under disturbance.

A. *Scenario 1: Typical working condition*

A typical working condition is used for investigating the effect of the prediction horizon length N, as shown in Fig. 4.14. The red line represents the bulldozer velocity (km/h) and the blue line represents the soil-cut depth (m). The operating periods are described as follows: the traveling stage is from 1–4 s; the soil-cutting stage is from 4–16 s; the soil-transportation stage is from 16–31 s; the unloading soil stage is from 31–33 s and the no-load stage is from 33–50 s. The required power calculated according to the equations described in Section 4.3 is shown in Fig. 4.15.

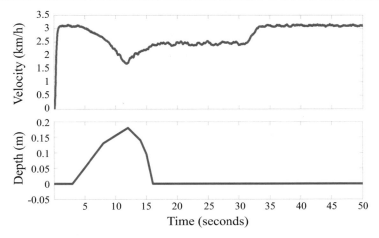

Figure 4.14: Typical working condition of HETB.

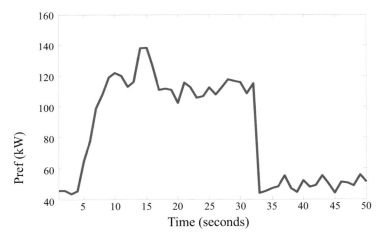

Figure 4.15: Power demand of the typical working condition.

The length of prediction horizon N is the most important MPC parameter that affects the solution. The prediction horizon N can be assigned to be 2 s, 4 s, or 15 s. Figure 4.16 shows the *SOE* profile corresponding to the different lengths assigned to N using MPC-based control strategy, and the optimal solution obtained from the DP algorithm. As expected, the MPC gets closer to the optimal solution from DP as the prediction horizon increases. The comparison of fuel consumption according to different prediction horizons is presented in Table 4.3. The fuel consumption also decreases with the increase of the receding horizon as seen from Table 4.3. For the reasons noted above, 15 s is chosen as the prediction horizon and applied in the MPC development under the two scenarios.

As the prediction horizon N is assigned to 15 s, the *SOE*, output power of engine, and the output power of UC is shown in Fig. 4.17. The trajectories of the out power of en-

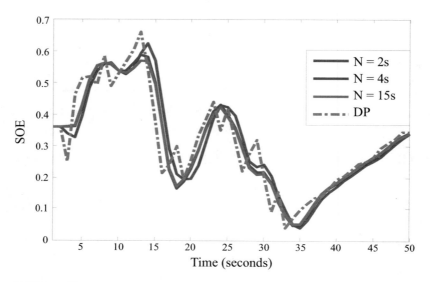

Figure 4.16: SOE profile with different lengths of prediction horizon.

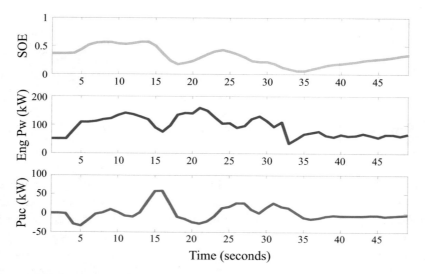

Figure 4.17: MPC results under scenario 1.

gine (represented by the blue line) and the out power of UC (represented by the pink line) demonstrate the optimal power split between two power resources to obtain the optimal fuel economy. The fuel consumption comparison calculated by three energy management strategies under this typical working condition is presented in Table 4.3. DP can obtain the minimum fuel consumption, 290 g, which can be the benchmark. The rule-based algorithm, from the previous work, is 313 g which can reach 92.6% of the optimal one. Compared with the rule-based strategy, an additional 6% of fuel economy is obtained by the MPC algorithm, and the MPC can achieve 98.6% fuel optimality of the DP. Although DP cannot be used in real time, analyzing its behavior can provide meaningful insight into the possible improvement of the MPC controller.

Table 4.3: Fuel consumption comparison under scenario 1

Control Strategy		Fuel Consumption (g)	Fuel Economy (%)
DP		290	100
Rule-based		313	92.6
MPC	N = 2	295.4	98.1
	N = 4	294.6	98.4
	N = 15	294	98.6

B. *Scenario 2: The Working Condition under Disturbances*

40% casual disturbances are added to the typical working condition to verify the robustness of the proposed MPC strategy, as shown in Fig. 4.18. The blue line is typical working conditon and the dotted line represents the disturbed working conditon.

The *SOE* of UC and the power distribution from two power sources under the second scenario is shown in Fig. 4.19. The blue line represents the output power of the engine and the red line represents the output power of the UC. The comparison of the *SOE* between the MPC and the DP under scenario 2 is shown in Fig. 4.20. Table 4.4 shows the comparison of fuel economy obtained from the three energy management strategies. Under the second scenario, the DP can obtain the optimal fuel economy for the benchmark. The MPC algorithm can obtain 98.9% fuel optimality of the DP benchmark under scenario 2 while the rule-based energy management can achieve 91%. We can make the conclusion that the MPC strategy is more effective when the working condition is not fully known.

4.3   HYBRID ENERGY STORAGE SYSTEM APPLICATION

With the high-speed development of hybrid electric road and off-road vehicles to deal with the serious air pollution and fossil fuel depletion problems [151, 152], energy storage systems

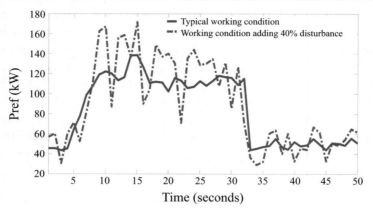

Figure 4.18: Power demand comparison under scenario 2.

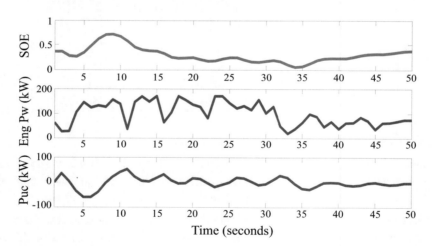

Figure 4.19: MPC results under scenario 2.

(ESS) play a key role in assisting the engine and storing regenerative braking energy in different types of HEVs. Generally, the ESS unit in HEVs should have the ability to provide both enough power and energy over different driving conditions. As the main electric energy storage systems, batteries and ultracapacitors have their own characteristics, as shown in Table 4.5. In addition, the battery life is much shorter than UC life, because the former is influenced by several factors, such as C-rate [153], depth of discharge (DOD) [154], high discharge current and high operating temperatures [155]. Batteries' attributes include low power density and high energy density. In addition, their degradation may accelerate when they experience high and frequent charg-

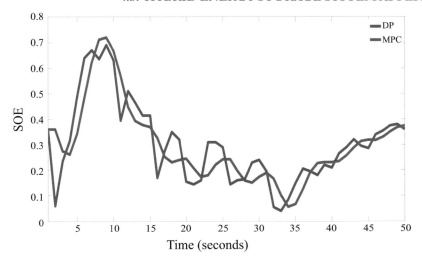

Figure 4.20: SOE profile comparison under scenario 2.

Table 4.4: Fuel consumption comparison in scenario 2

Control Strategy	Fuel Consumption (g)	Fuel Economy (%)
DP	304.7	100
Rule-based	334.8	91
MPC	308	98.9

ing/discharging [153, 154, 155]. As a complementary ESS, the UC has the advantages of high power density, a long lifespan, high efficiency as well as a wide range of operating temperatures.

The HESS combines the UC and battery can get a better overall performance than either the UC or the battery alone, which has attracted a large amount of attention recently [156, 157].

Off-road HEVs, such as hybrid electric tracked bulldozers (HETB), are different from road HEVs. Unlike the road HEVs, the auxiliary device in off-road HEVs (e.g., bulldozer) consumes a large portion of the total power, which cannot be neglected. In addition, the auxiliary device should be able to activate very quickly. Therefore, an ultracapacitor seems more suitable than a battery pack since the UC can quickly provide large amounts of power without significantly impacting its life span. Based on such analysis, all of the reported hybrid bulldozers, such as the D7E HETB produced by Caterpillar in the U.S. and Shantui Construction Machinery Co., Ltd in China shown in our previous research work [147], adopt the UC as the sole ESS. However, the main disadvantages of only using UC in HETB are their significant total cost, along with the significantly added volume and weight. Furthermore, by analyzing a typical working condition of an HETB, it can be seen that about half of the whole cycle still consumes a constant power of 50 kw. According to the aforementioned analysis, it is reasonable and beneficial to introduce the HESS to HETB, which combines the battery and UC. The powertrain of the proposed HETB is shown in Fig. 4.21. The HESS is composed of a combined battery and UC pack, and the assistance power unit (APU) system includes the engine and generator.

Table 4.5: Typical properties of the battery and ultracapacitor

Chemistry	Power Density (kW/kg)	Energy Density (Wh/kg)	Cycle Life at 80% DOD
Lead Acid	0.18	30–40	~800
Ni-Mh Battery	0.4–1.2	55–80	~2,000
Li-Ion Battery	0.8–2	80–170	~3,000
Li-Polymer Battery	1–2.8	130–200	~2,000
Ultracapacitor	4–10	2–30	>1,000,000

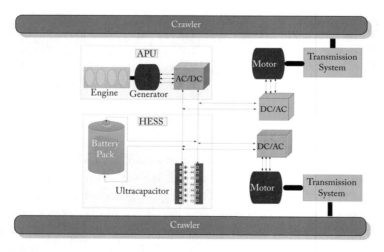

Figure 4.21: Powertrain of hybrid tracked bulldozer.

4.3.1   MODELING OF OFF-ROAD VEHICLE WITH HESS

In the HETB, the model's vehicle, engine, generator, motor, and UC adopt the same models as Section 3.1 Modeling of series hybrid bulldozer. The only new model here is the battery model which is demonstrated as follow.

The equivalent resistance model is adopted. In this model, V_{bat} denotes the open-circuit voltage; R_b denotes the equivalent resistance; I_{bat} denotes the battery output current; U_{bat} denotes the battery terminal voltage.

$$U_{bat} = V_{bat} - I_{bat}R_b \qquad (4.17)$$

Both the equivalent resistance R_b and the open-circuit voltage V_{bat} vary according to the changes of the temperature and SOC; however, the effect of temperature is small and can be neglected when modeling. V_{bat} and R_b are obtained through lookup tables by inputting the SOC, and the SOC of the battery pack is calculated by:

$$SOC(t) = \frac{Q_0 - \int_0^t i(\tau)d\tau}{Q_{\max}} \qquad (4.18)$$

where Q_0 is the initial capacity of the battery; $SOC(t)$ is the battery's SOC of the current time; $i(\tau)$ is the battery's charge-discharge current; and $Q_{\max}$ is the maximum capacity of the battery. The dynamic equation of the battery SOC variation is shown as:

$$SOC(t) = \begin{cases} -\frac{1}{\eta_{batt}}\frac{P_{batt}(t)}{Q_{\max}V_{batt}} & \text{if } P_{batt}(t) \geq 0 \text{ (discharge)} \\ -\eta_{batt}\frac{P_{batt}(t)}{Q_{\max}V_{batt}} & \text{if } P_{batt}(t) < 0 \text{ (charge)} \end{cases} \qquad (4.19)$$

where η_{batt} is the battery's efficiency.

The main battery cell parameters (Lithium Ion ANR26650 MI) selected in this study are reported in Table 4.7. In this case, the battery pack is comprised of 91 battery cells combined with the UC pack illustrated in Table 4.6 to formulate the HESS.

Table 4.6: Ultracapacitor pack parameters

Description	Quantity	Unit
Nominal capacity	4.8	[F]
Nominal voltage	300	[V]
Peak cell discharging current	100	[A]
Weight	130	[kg]
Volume	300	[mm]
Number of Cycles	>50,000	—

In order to evaluate the battery degradation phenomenon, a phenomenological battery life [160] estimation method is adopted, which is based on a damage accumulation model and the concept of accumulated Ah-throughput, i.e., the total amount of charge that can flow through the battery before it reaches the end of life. The Ah-throughput is computed to evaluate

Table 4.7: Battery cell parameters

Description	Quantity	Unit
Nominal cell capacity	2.3	[Ah]
Nominal cell voltage	3.3	[V]
Peak cell discharging current	100	[A]
Weight	70	[g]
Volume	Ø26*65	[mm]
Resistance	10	[mΩ]
Number of Cycles	>1,000	—

the actual depletion of the battery charge:

$$Ah_{eff}(t) = \int_0^t \sigma\left(I_{bat}(\tau),\, T_{bat}(\tau),\, SOC(\tau)\right)|I_{bat}(\tau)|\,d\tau \tag{4.20}$$

which represents the amount of charge that would need to go through the battery using the nominal cycle to have the same aging effect of the actual conditions. When the total Ah-throughput under a nominal drive cycle, as defined by the manufacturer (1-C rate, 25°, 100% DOD), becomes Γ, the end of the battery life is reached.

Where:

$$\Gamma = \int_0^{EOL} |I_{nom}(t)|\,dt \tag{4.21}$$

and EOL indicates the end of life and I_{nom} denotes the nominal current. For a given battery, the quantity of Γ is a constant, and $\Gamma = 20{,}000$ Ah in this case.

This model relies on the severity factor σ as shown in (4.20). At the cell level, the severity factor depends on the C-rate, the temperature T and the Depth of Discharge (DOD) [160]. C-rate refers to the discharge intensity of a battery, and battery life can diminish when cycling at high C-rates [158, 159]. In order to analyze the HESS Performance, the battery C-rate is denoted by [161],

$$C - rate = \frac{P_{batt}}{0.69 V_{bat} Q_{\max}} \tag{4.22}$$

Under a given drive cycle, the severity factor is defined as:

$$\sigma = \frac{\gamma\left(I_{bat}, T, SOC\right)}{\Gamma} = \frac{\int_0^{EOL} |I_{bat}|\,dt}{\int_0^{EOL} |I_{nom}|\,dt} \tag{4.23}$$

where $\gamma(I_{bat}, T, SOC)$ is the battery duration (Ah-throughout) with a given sequence of current, temperature, and SOC. The severity factor represents the relative aging effect under the nominal cycle, and it is higher than 1 when the conditions are more severe in terms of aging. This factor should be identified by aging experiments. The published data from [161] are used to create the severity factor map shown in Fig. 4.22; whereas, the data for Li-Ion batteries are from A123Systems [162]. High C-rate, low operating temperatures, and low values of *SOC* significantly contribute to the increase of the severity factor. Many researchers have analyzed the HESS performance in order to extend battery life by the accumulated Ah-throughput model [163].

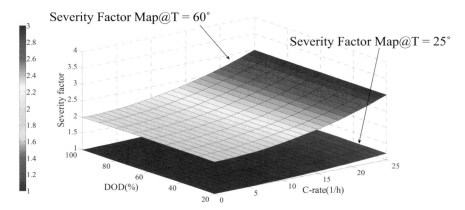

Figure 4.22: Battery severity factor map.

4.3.2 ENERGY MANAGEMENT FOR HYBRID OFF-ROAD VEHICLE WITH HESS

An MPC based strategy is adopted to deal with the energy management of off-road vehicles. An MPC-based energy management strategy is presented to coordinate the power flow between different ESS. It is designed based on the total cost of the whole system by considering downsizing the UC and improving the battery lifespan. Furthermore, the performance and total cost of the HETB with HESS is quantitatively studied and compared to the ones with only a UC and only a battery pack.

According to the HETB model developed in the previous section, the MPC can be developed using the following equations:

$$x_1 = \begin{cases} -\frac{1}{\eta_{batt}} \frac{P_{batt}(t)}{Q_{max} V_{batt}} & \text{if } P_{batt}(t) \geq 0 \quad \text{(discharge)} \\ -\eta_{batt} \frac{P_{batt}(t)}{Q_{max} V_{batt}} & \text{if } P_{batt}(t) < 0 \quad \text{(charge)} \end{cases}$$

$$x_2 = \begin{cases} -\frac{1}{\eta_{cap}} \frac{P_{uc}(t)}{E_{cap}} & \text{if } P_{uc}(t) \geq 0 \quad \text{(discharge)} \\ -\eta_{cap} \frac{P_{uc}(t)}{E_{cap}} & \text{if } P_{uc}(t) < 0 \quad \text{(charge)} \end{cases} \qquad (4.24)$$

$$x_3 = B_e (P_e) = B_e \left(\frac{P_{req} - P_{batt} - P_{uc}}{\eta_g} \right)$$

where $x_1 = SOC$, $x_2 = SOE$, and $x_3 = B_e$ denotes the fuel consumption; $u = P_{uc}$ represents the control input.

A state-space representation of the system has been implemented in the MPC algorithm. The vectors of states, control inputs, measured inputs, as well as the outputs are defined in (4.25):

$$x = \begin{bmatrix} SOC \\ SOE \\ B_e \end{bmatrix}, \quad u = \begin{bmatrix} P_{batt} \\ P_{uc} \end{bmatrix}, \quad v = P_{req}, \quad y = \begin{bmatrix} SOC \\ SOE \\ B_e \end{bmatrix} \qquad (4.25)$$

The linearized and discretized model of the system becomes:

$$\begin{cases} x(k+1) = A(k)x(k) + B_u(k)u(k) + B_v(k) \\ y(k) = C(k)x(k) \end{cases} \qquad (4.26)$$

In this equation,

$$A = \begin{bmatrix} 1 & 0 & 0 \\ 0 & 1 & 0 \\ 0 & 0 & 1 \end{bmatrix}; \quad B_u(k) = \begin{bmatrix} -\frac{1}{Q_{batt}*2160} & 0 \\ 0 & -\frac{1}{E_{cap}} \\ -m_1 & -m_1 \end{bmatrix};$$

$$B_v(k) = \begin{bmatrix} 0 \\ 0 \\ m_1 * P_{ref} + m_2 \end{bmatrix}; \quad C(k) = \begin{bmatrix} 1 & 0 & 0 \\ 0 & 1 & 0 \\ 0 & 0 & 1 \end{bmatrix}$$

To analyze the performance of the proposed HESS, the fuel consumption, the battery C-rate, and variation of the battery power are considered in the cost function. The cost function can be described by:

$$J(k) = \sum_{i=1}^{N-1} \left[\|Y(i) - Y_{ref}(i)\|_Q^2 + \|C - rate(i)\|_{R_1}^2 + \|\Delta P_{batt}(i)\|_{R_2}^2 \right] \qquad (4.27)$$

where, i indicates a sample of the prediction; $Y(i)$ is the actual output of the system; $Y_{ref}(i)$ is the reference output of the system; and Q, R_1, R_2 are the weighting matrices:

$$Q = \begin{bmatrix} 100000000 & 0 & 0 \\ 0 & 10000000 & 0 \\ 0 & 0 & 10 \end{bmatrix}; \quad R_1 = \begin{bmatrix} 10000000 & 0 \\ 0 & 0 \end{bmatrix}; \quad R_2 = \begin{bmatrix} 100000 & 0 \\ 0 & 0 \end{bmatrix}$$

The physical constraints of state and control variables are denoted by the following inequalities to guarantee smooth/safe operation of the key components, including the engine, motor, battery, and UC:

$$\begin{cases} SOC_{\min} \le SOC \le SOC_{\max}; \\ SOE_{\min} \le SOE \le SOE_{\max}; \\ N_{e\_\min} \le N_e \le N_{e\_\max}; \\ P_{e\_\min} \le P_e \le P_{e\_\max}; \\ T_{e\_\min} \le T_e \le T_{e\_\max} \end{cases} \tag{4.28}$$

The energy management problem is solved by an open source solver, qpOASES [164]. After the optimal control input sequence $u_0, u_1, u_2 \ldots u_{N-1}$ is obtained, the first element of this trajectory u_0 will be applied to the plant model of the HETB. The updated value of the state is obtained in the subsequent step. The receding control strategy is implemented by repeating this procedure during subsequent time steps.

The working scenario consists of nine disturbance-added drive cycles as shown in Fig. 4.15 as the combined typical working condition.

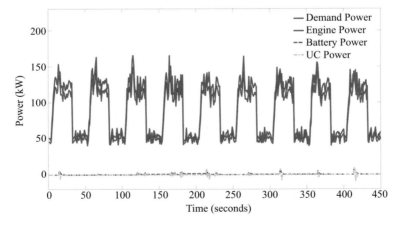

Figure 4.23: Power distribution.

Figure 4.23 illustrates the power distribution between engine, battery, and UC to meet the power demand. The majority of the power is provided from the engine, and the battery and UC provide the power requirements exceeding the power limit of the engine. The distribution of

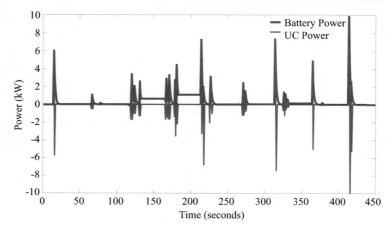

Figure 4.24: UC power and battery power.

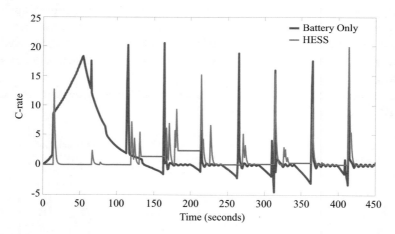

Figure 4.25: C-rate comparison between battery only and HESS.

the battery and UC power is reported in Fig. 4.24. As shown in Fig. 4.24, the sharply changing power is provided by the UC pack, and the relatively smooth power is provided by the battery, which decreases the damage to the battery.

The C-rate comparison between the HESS and battery only is reported in Fig. 4.25. It can be observed that the maximum and average C-rate of the battery are both improved by using the HESS, compared to using battery only. Furthermore, the fluctuation of the C-rate is smoother after using HESS, with the average C-rate reduced by 60.3%. As noted previously, the C-rate directly impacts the battery life—therefore, the improved C-rate of the battery is expected to extend battery life and reliability.

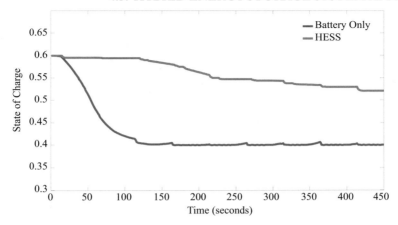

Figure 4.26: Battery state of charge.

The battery state-of-charge comparison between battery only and HESS is reported in Fig. 4.26. The initial state of charge is set at 60%. The plot shows that the HESS maintains a higher SOC compared to the battery only and the fluctuation of the SOC gets smoother due to the UC getting involved in the ESS. For battery only, the final SOC reached 0.4; meanwhile, the final SOC reached 0.5205 for the HESS, as shown in Table 4.8. The RMS (root mean square) of the battery current is decreased by 59.7% compared with the HESS.

The fuel economy improvement of the three kinds of ESSs compared with the conventional bulldozer is reported in Table 4.8. The total fuel consumption of these three ESSs is very close and the bulldozer equipped with the UC pack only can achieve the best fuel economy by 22% compared with the conventional bulldozer. The bulldozer equipped with battery only achieves 21.37% fuel economy improvement, which is similar to the HESS which improves the fuel economy by 21.89%.

In order to predict the battery change time and the ESS costs of the bulldozer in its whole life years, we assume that the working hours of the bulldozer per day are eight hours, and that it is used 2,500 days for its whole life (ten years). As shown in Table 4.8, the Aheff calculated by (11) is 0.1025 Ah when using battery only, and is 0.0254 Ah when equipped with HESS. The Aheff calculated by (11) in this scenario is reduced by 75.2%. Therefore, we can conclude that, under scenario 1, the battery can be used for 457 days and will be changed 5.4 times using battery only, whereas the HESS will be used for 1,875 days and will be changed just 1.6 times before it is scrapped. The predicted battery life of HESS can be extended by 310% compared with battery only. As shown in Fig. 4.27, the ESS costs will be decreased by 36% and 32.5% compared with battery only and UC only. This data shows the financial viability of HESS. In addition to the cost benefits from adopting HESS in this off-road vehicle, the ESS volume

Table 4.8: Performance comparison in scenario 1

Powertrain	Engine Only	Battery Only	UC Only	HESS
Final SOC	—	0.400	—	0.5205
Final SOE	—	—	0.5437	0.3277
Total consumption (g)	3435.35	2701.03	2678	2683.3
Fuel economy improvement (%)	—	21.37	22	21.89
Average C-rate	—	2.32	—	0.92
RMS battery current (A)	—	8.98	—	3.62
Ah_{eff} (Ah)	—	0.1025	—	0.0254
Battery change times	—	5.4	—	1.6

and the weight are also reduced significantly by adopting HESS. The weight can be reduced by 40.77% compared with UC only, and the volume can be downsized by nearly 50%.

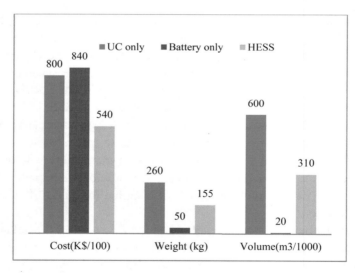

Figure 4.27: ESS performance comparison chart.

References

[1] https://www.infineon.com/cms/en/applications/commercial-construction-and-agricultural-vehicles/hybrid-electric-solutions-for-construction-commercial-and-agricultural-vehicles/ 3, 21

[2] Y. Huang, A. Khajepour, T. Zhu, and H. Ding, A Supervisory Energy-saving controller for a novel anti-idling system of service vehicles, *IEEE/ASME Transactions on Mechatronics*, vol. 22, no. 2, pp. 1037–1046, 2017. DOI: 10.1109/tmech.2016.2631897. 3, 21

[3] Y. Huang, H. Wang, A. Khajepour, H. He, and J. Ji, Model predictive control power management strategies for HEVs: A review, *Journal of Power Sources*, vol. 341, pp. 91–106, 2017. DOI: 10.1016/j.jpowsour.2016.11.106. 3, 21, 44

[4] H. Wang, Y. Huang, A. Khajepour, and Q. Song, Model predictive control-based energy management strategy for a series hybrid electric tracked vehicle, *Applied Energy*, vol. 182, pp. 105–114, 2016. DOI: 10.1016/j.apenergy.2016.08.085. 4, 21

[5] http://www.heavyequipmentguide.ca/article/24299/volvo-construction-equipment-unveils-prototype-hybrid-wheel-loader-the-lx1 4, 8, 21

[6] http://www.rockproducts.com/technology/loading-a-hauling/12211-ease-of-operation-fuel-efficiency-drive-john-deere-644k-hybrid-loader.html#.WXpLUIjyu71 21

[7] http://articles.sae.org/14803/ 21

[8] http://articles.sae.org/2370/ 22

[9] F. R. Salmasi, Control strategies for hybrid electric vehicles: Evolution, classification, comparison, and future trends, *IEEE Transactions on Vehicular Technology*, vol. 56, no. 5, pp. 2393–2404, September 2007. DOI: 10.1109/tvt.2007.899933. 34

[10] S. F. Tie and C. W. Tan, A review of energy sources and energy management system in electric vehicles, *Renewable and Sustainable Energy Reviews*, vol. 20, pp. 82–102, April 2013. DOI: 10.1016/j.rser.2012.11.077. 34, 35, 36, 37

[11] M. Kazmierkowski, Electric power systems (review of modern electric, hybrid electric, and fuel cell vehicles: Fundamentals, theory, and design, 2nd ed., Ehsani, Y. G. And Emadi, A., 2010) [Book News], *IEEE Industrial Electronics Magazine*, vol. 4, no. 1, pp. 75–75, March 2010. DOI: 10.1109/mie.2010.936103. 34, 37

[12] S. Jeon, S. Jo, Y. Park, and J. Lee, Multi-mode driving control of a parallel hybrid electric vehicle using driving pattern recognition, *Journal of Dynamic Systems, Measurement, and Control*, vol. 124, no. 1, p. 141, 2002. DOI: 10.1115/1.1434264. 22, 34

[13] Y. Yasuhiro, et al., Driving pattern prediction for an energy management system of hybrid electric vehicles in a specific driving course, *Industrial Electronics Society, IECON. 30th Annual Conference of IEEE*, vol. 2, 2004. DOI: 10.1109/iecon.2004.1431842. 34

[14] K. T. Chau and Y. S. Wong, Overview of power management in hybrid electric vehicles, *Energy Conversion and Management*, vol. 43, no. 15, pp. 1953–1968, October 2002. DOI: 10.1016/s0196-8904(01)00148-0. 34

[15] V. Johnson, K. B. Wipke, and D. J. Rausen, HEV control strategy for real-time optimization of fuel economy and emissions, *SAE Transactions* 109.3(2000):1677–1690. DOI: 10.4271/2000-01-1543. 35

[16] A. M. Phillips, M. Jankovic, and K. E. Bailey, Vehicle system controller design for a hybrid electric vehicle, *Control Applications, Proceedings of the IEEE International Conference on*, 2000. DOI: 10.1109/cca.2000.897440. 35

[17] R. Pusca, J. Kauffmann, A. Berthon, and Y. Ait-Amirat, Fuzzy-logic-based control applied to a hybrid electric vehicle with four separate wheel drives, *IEE Proc.—Control Theory and Applications*, vol. 151, no. 1, pp. 73–81, January 2004. DOI: 10.1049/ip-cta:20040066. 35

[18] F. U. Syed, M. L. Kuang, M. Smith, S. Okubo, and H. Ying, Fuzzy gain-scheduling proportional—integral control for improving engine power and speed behavior in a hybrid electric vehicle, *IEEE Transactions on Vehicular Technology*, vol. 58, no. 1, pp. 69–84, January 2009. DOI: 10.1109/tvt.2008.923690. 35

[19] A. Sciarretta and L. Guzzella, Control of hybrid electric vehicles, *IEEE Control Systems Magazine*, vol. 27, no. 2, pp. 60–70, 2007. DOI: 10.1109/mcs.2007.338280. 35

[20] W. Xiong, Y. Zhang, and C. Yin, Optimal energy management for a series—parallel hybrid electric bus, *Energy Conversion and Management*, vol. 50, no. 7, pp. 1730–1738, July 2009. DOI: 10.1016/j.enconman.2009.03.015. 36

[21] S. Gang, et al., Study and simulation of based-fuzzy-logic parallel hybrid electric vehicles control strategy, *6th International Conference on Intelligent Systems Design and Applications*, vol. 1, IEEE, 2006. DOI: 10.1109/isda.2006.252. 36

[22] H.-D. Lee et al., Torque control strategy for a parallel-hybrid vehicle using fuzzy logic, *IEEE Industry Applications Magazine*, vol. 6, no. 6, pp. 33–38, 2000. DOI: 10.1109/2943.877838. 36

[23] J.-S. Won and R. Langari, Fuzzy torque distribution control for a parallel hybrid vehicle, *Expert Systems*, vol. 19, no. 1, pp. 4–10, February 2002. DOI: 10.1111/1468-0394.00184. 36

[24] B. M. Baumann, G. Washington, B. C. Glenn, and G. Rizzoni, Mechatronic design and control of hybrid electric vehicles, *IEEE/ASME Transactions on Mechatronics*, vol. 5, no. 1, pp. 58–72, March 2000. DOI: 10.1109/3516.828590. 36

[25] N. J. Schouten, M. A. Salman, and N. A. Kheir, Fuzzy logic control for parallel hybrid vehicles, *IEEE Transactions on Control Systems Technology*, vol. 10, no. 3, pp. 460–468, May 2002. DOI: 10.1109/87.998036. 36

[26] M. Mohebbi, M. Charkhgard, and M. Farrokhi, Optimal neuro-fuzzy control of parallel hybrid electric vehicles, *IEEE Vehicle Power and Propulsion Conference*, IEEE, 2005. DOI: 10.1109/vppc.2005.1554566. 36

[27] C.ZhiHang, M. A. Masrur, and Y. L. Murphey, Intelligent vehicle power management using machine learning and fuzzy logic, *Fuzzy Systems, FUZZ-IEEE. (IEEE World Congress on Computational Intelligence). IEEE International Conference on*, 2008. DOI: 10.1109/fuzzy.2008.4630697. 36

[28] E. D. Tate and S. P. Boyd, Finding ultimate limits of performance for hybrid electric vehicles, *No. 2000–01–3099. SAE Technical Paper*, 2000. DOI: 10.4271/2000-01-3099. 37

[29] L. Chan-Chiao, et al., Power management strategy for a parallel hybrid electric truck, *IEEE Transactions on Control Systems Technology*, 11.6(2003):839–849. DOI: 10.1109/tcst.2003.815606. 37

[30] Z. Chen, C. C. Mi, R. Xiong, J. Xu, and C. You, Energy management of a power-split plug-in hybrid electric vehicle based on genetic algorithm and quadratic programming, *Journal of Power Sources*, vol. 248, pp. 416–426, February 2014. DOI: 10.1016/j.jpowsour.2013.09.085. 37

[31] H. Hongwen, et al., Energy management strategy research on a hybrid power system by hardware-in-loop experiments, *Applied Energy*, 112(2013):1311–1317. DOI: 10.1016/j.apenergy.2012.12.029. 37

[32] X. M. Wang, H. W. He, F. C. Sun, J. L. Zhang, Application study on the dynamic programming algorithm for energy management of plug-in hybrid electric vehicles, *Energies*, vol. 8, pp. 3225–3244, 2015. DOI: 10.3390/en8043225. 37

[33] S. Kermani, S. Delprat, T. M. Guerra, R. Trigui, and B. Jeanneret, Predictive energy management for hybrid vehicle, *Control Engineering Practice*, vol. 20, pp. 408–420, April 2012. DOI: 10.1016/j.conengprac.2011.12.001. 37

[34] J. Ribau, R. Viegas, A. Angelino, A. Moutinho, and C. Sliva, A new offline optimization approach for designing a fuel cell hybrid bus, *Transactions on Research Part C: Emerging Technologies*, vol. 42, pp. 14–25, May 2014. DOI: 10.1016/j.trc.2014.02.012. 37

[35] H. W. He, H. L. Tang, and X. M. Wang, Global optimal energy management strategy research for a plug-in series-parallel hybrid electric bus by using dynamic programming, *Mathematical Problems Engineering*, vol. 2013. DOI: 10.1155/2013/708261. 37

[36] A. Brahma, Y. Guezennec, and G. Rizzoni, Optimal energy management in series hybrid electric vehicles, *American Control Conference, Proceedings of the*, vol. 1, no. 6, IEEE, 2000. DOI: 10.1109/acc.2000.878772. 37

[37] B. Huang, Z. Wang, and Y. Xu, Multi-objective genetic algorithm for hybrid electric vehicle parameter optimization, *Intelligent Robots and Systems, IEEE/RSJ International Conference on*, 2006. DOI: 10.1109/iros.2006.281654. 39

[38] L.-C. Fang and S.-Y. Qin, Concurrent optimization for parameters of power-train and control system of hybrid electric vehicle based on multi-objective genetic algorithms, *SICE-ICASE, International Joint Conference*, IEEE, 2006. DOI: 10.1109/sice.2006.315114. 39

[39] M. Montazeri-Gh, A. Poursamad, and B. Ghalichi, Application of genetic algorithm for optimization of control strategy in parallel hybrid electric vehicles, *Journal of the Franklin Institute*, 343.4(2006):420–435. DOI: 10.1016/j.jfranklin.2006.02.015. 39

[40] M. Montazeri-Gh, A. Ahmadi, and M. Asadi, Driving condition recognition for genetic-fuzzy HEV control, *Genetic and Evolving Systems, GEFS, 3rd International Workshop on*, IEEE, 2008. DOI: 10.1109/gefs.2008.4484569. 39

[41] R. S. Wimalendra, et al., Determination of maximum possible fuel economy of HEV for known drive cycle: Genetic algorithm based approach, *Information and Automation for Sustainability, ICIAFS, 4th International Conference on*, IEEE, 2008. DOI: 10.1109/ici-afs.2008.4783975. 39

[42] T. Yi, Z. Xin, and Z. Liang, Fuzzy-genetic control strategy of hybrid electric vehicle, *Intelligent Computation Technology and Automation, ICICTA'09, 2nd International Conference on*, vol. 2, IEEE, 2009. DOI: 10.1109/icicta.2009.409. 39

[43] C. Desai and S. S. Williamson, Optimal design of a parallel hybrid electric vehicle using multi-objective genetic algorithms, *Vehicle Power and Propulsion Conference, VPPC'09*, IEEE, 2009. DOI: 10.1109/vppc.2009.5289754. 39

[44] C. Desai and S. S. Williamson, Optimal design of a parallel hybrid electric vehicle using multi-objective genetic algorithms, *Vehicle Power and Propulsion Conference, VPPC'09*, IEEE, 2009. DOI: 10.1109/vppc.2009.5289754. 39

[45] G. Paganelli, T. M. Guerra, S. Delprat, J.-J. Santin, M. Delhom, and E. Combes, Simulation and assessment of power control strategies for a parallel hybrid car, *Proc. of the Institution of Mechanical Engineers, Part D: Journal of Automobile Engineering*, vol. 214, no. 7, pp. 705–717, January 2000. DOI: 10.1243/0954407001527583. 39

[46] G. Paganelli, et al., Control development for a hybrid-electric sport-utility vehicle: Strategy, implementation and field test results, *American Control Conference, Proceedings of the*, vol. 6, IEEE, 2001. DOI: 10.1109/acc.2001.945787. 39

[47] G. Paganelli, et al., Equivalent consumption minimization strategy for parallel hybrid powertrains, *Vehicular Technology Conference, VTC Spring, IEEE 55th*, vol. 4, 2002. DOI: 10.1109/vtc.2002.1002989. 39

[48] N. Cui, et al., Research on predictive control based energy management strategy for hybrid electric vehicle, *Power Electronics for Distributed Generation Systems (PEDG), 3rd IEEE International Symposium on*, 2012. DOI: 10.1109/pedg.2012.6254070. 40

[49] P. Tulpule, V. Marano, and G. Rizzoni, Effects of different PHEV control strategies on vehicle performance, *American Control Conference, ACC'09*, IEEE, 2009. DOI: 10.1109/acc.2009.5160595. 40

[50] P. Pisu and G. Rizzoni, A supervisory control strategy for series hybrid electric vehicles with two energy storage systems, *Vehicle Power and Propulsion, IEEE Conference*, 2005. DOI: 10.1109/vppc.2005.1554534. 40

[51] B. Gu and G. Rizzoni, An adaptive algorithm for hybrid electric vehicle energy management based on driving pattern recognition, *ASME International Mechanical Engineering Congress and Exposition*, American Society of Mechanical Engineers, 2006. DOI: 10.1115/imece2006-13951. 40

[52] G. Paganelli, et al., General supervisory control policy for the energy optimization of charge-sustaining hybrid electric vehicles, *JSAE Review*, 22.4(2001):511–518. DOI: 10.1016/s0389-4304(01)00138-2. 40

[53] G. Paganelli, et al., Equivalent consumption minimization strategy for parallel hybrid powertrains, *Vehicular Technology Conference, VTC Spring, IEEE 55th*, vol. 4, 2002. DOI: 10.1109/vtc.2002.1002989. 40

[54] J. G. Supina, and S. Awad, Optimization of the fuel economy of a hybrid electric vehicle, *Circuits and Systems, IEEE 46th Midwest Symposium on*, vol. 2, 2003. DOI: 10.1109/mwscas.2003.1562426. 40

[55] A. Sciarretta, M. Back, and L. Guzzella, Optimal control of parallel hybrid electric vehicles, *IEEE Transactions on Control Systems Technology*, 12.3(2004):352–363. DOI: 10.1109/tcst.2004.824312. 40

[56] J.-S. Won, R. Langari, and M. Ehsani, An energy management and charge sustaining strategy for a parallel hybrid vehicle with CVT, *IEEE Transactions on Control Systems Technology*, 13.2(2005):313–320. DOI: 10.1109/tcst.2004.838569. 40

[57] P. Pisu, et al., A LMI-based supervisory robust control for hybrid vehicles, *American Control Conference, Proceedings of the*, vol. 6, IEEE, 2003. DOI: 10.1109/acc.2003.1242462. 40

[58] P. Pisu, and G. Rizzoni, H∞ control for hybrid electric vehicles, *Decision and Control, CDC. 43rd IEEE Conference on*, vol. 4, 2004. DOI: 10.1109/CDC.2004.1429253. 40

[59] Y. L. Murphey, J. Park, Z. Chen, M. L. Kuang, M. A. Masrur, and A. M. Phillips, Intelligent hybrid vehicle power Control—Part I: Machine learning of optimal vehicle power, *IEEE Transactions on Vehicular Technology*, vol. 61, no. 8, pp. 3519–3530, October 2012. DOI: 10.1109/tvt.2012.2206064. 40, 46

[60] R. Liu, D. Shi, and C. Ma, Real-time control strategy of Elman neural network for the parallel hybrid electric vehicle, *Journal of Applied Mathematics*, vol. 2014, pp. 1–11, 2014. DOI: 10.1155/2014/596326. 40

[61] Z. Wang, W. Li, and Y. Xu, A novel power control strategy of series hybrid electric vehicle, *IEEE/RSJ International Conference on Intelligent Robots and Systems*, 2007. DOI: 10.1109/iros.2007.4399024. 40

[62] H. H. Chin and A. A. Jafari, A selection algorithm for power controller unit of hybrid vehicles, *14th International IEEE Conference on Intelligent Transportation Systems (ITSC)*, 2011. DOI: 10.1109/itsc.2011.6082910. 41

[63] J. Park, et al., Intelligent vehicle power control based on machine learning of optimal control parameters and prediction of road type and traffic congestion, *IEEE Transactions on Vehicular Technology*, 58.9(2009):4741–4756. DOI: 10.1109/tvt.2009.2027710. 41

[64] N. Giorgetti, G. Ripaccioli, A. Bemporad, I. V. Kolmanovsky, D. Hrovat, Hybrid model predictive control of direct injection stratified charge engines, *IEEE/ASME Transactions on Mechatronics*, 11(5), pp. 499–506, 2006. DOI: 10.1109/tmech.2006.882979. 41

[65] F. A. Bender, M. Kaszynski, and O. Sawodny, Drive cycle prediction and energy management optimization for hybrid hydraulic vehicles, *IEEE Transactions on Vehicular Technology*, vol. 62, no. 8, pp. 3581–3592, October 2013. DOI: 10.1109/tvt.2013.2259645. 48

[66] F. Yan, J. Wang, and K. Huang, Hybrid electric vehicle model predictive control torque-split strategy incorporating engine transient characteristics, *Vehicular Technology, IEEE Transactions on*, 61(6), pp. 2458–2467. DOI: 10.1109/tvt.2012.2197767. 41

[67] D. F. Opila, X. Wang, R. McGee, and J. W. Grizzle, Real-time implementation and hardware testing of a hybrid vehicle energy management controller based on stochastic dynamic programming, *Journal of Dynamic Systems, Measurement, and Control*, vol. 135, no. 2, p. 021002, November 2012. DOI: 10.1115/1.4007238. 44

[68] J. Tang, et al., Energy management of a parallel hybrid electric vehicle with CVT using model predictive control, *Control Conference (CCC), 35th Chinese. TCCT*, 2016. DOI: 10.1109/chicc.2016.7554036. 41

[69] D. Rotenberg, A. Vahidi, and I. Kolmanovsky, Ultracapacitor assisted Power-trains: Modeling, control, sizing, and the impact on fuel economy, *IEEE Transactions on Control Systems Technology*, vol. 19, no. 3, pp. 576–589, May 2011. DOI: 10.1109/tcst.2010.2048431. 41

[70] H. Borhan, A. Vahidi, A. M. Phillips, M. L. Kuang, I. V. Kolmanovsky, and S. Di Cairano, MPC-based energy management of a power-split hybrid electric vehicle, *IEEE Transactions on Control Systems Technology*, vol. 20, no. 3, pp. 593–603, May 2012. DOI: 10.1109/tcst.2011.2134852. 41

[71] V. Ngo, et al., Predictive gear shift control for a parallel hybrid electric vehicle, *IEEE Vehicle Power and Propulsion Conference*, 2011. DOI: 10.1109/vppc.2011.6043185. 41

[72] L. Johannesson, M. Asbogard, and B. Egardt, Assessing the potential of predictive control for hybrid vehicle Powertrains using stochastic dynamic programming, *IEEE Transactions on Intelligent Transportation Systems*, vol. 8, no. 1, pp. 71–83, March 2007. DOI: 10.1109/itsc.2005.1520076. 41

[73] R. Wang and S. M Lukic, Review of driving conditions prediction and driving style recognition based control algorithms for hybrid electric vehicles, *IEEE Vehicle Power and Propulsion Conference*, pp. 1–7. DOI: 10.1109/vppc.2011.6043061. 42

[74] S. Di Cairano, D. Bernardini, A. Bemporad, and I. V. Kolmanovsky, Stochastic MPC with learning for driver-predictive vehicle control and its application to HEV energy management, *IEEE Transactions on Control Systems Technology*, vol. 22, no. 3, pp. 1018–1031, May 2014. DOI: 10.1109/tcst.2013.2272179. 42, 43, 45

[75] H. Banvait, J. Hu and Y. Chen, Energy management control of plug-in hybrid electric vehicle using hybrid dynamical systems, *IEEE Transactions on Intelligent transportation systems*, 2013. 42, 43

[76] H. Wang, Y. Huang, A. Khajepour, and Q. Song, Model predictive control-based energy management strategy for a series hybrid electric tracked vehicle, *Applied Energy*, 182(2016):105–114. DOI: 10.1016/j.apenergy.2016.08.085. 43

[77] S. Fekri and F. Assadian, Fast model predictive control and its application to energy management of hybrid electric vehicles, *INTECH Open Access Publisher*, 2011. DOI: 10.5772/16319. 43

[78] A. Santucci, A. Sorniotti, and C. Lekakou, Power split strategies for hybrid energy storage systems for vehicular applications, *Journal of Power Sources*, vol. 258, pp. 395–407, July 2014. DOI: 10.1016/j.jpowsour.2014.01.118. 43

[79] H. A. Borhan, A. Vahidi, A. M. Phillips, M. L. Kuang, and I. V. Kolmanovsky, Predictive energy management of a power-split hybrid electric vehicle, *American Control Conference*, pp. 3970–3976, IEEE, June 10, 2009. DOI: 10.1109/acc.2009.5160451. 43

[80] S. Zhang, R. Xiong, and F. Sun, Model predictive control for power management in a plug-in hybrid electric vehicle with a hybrid energy storage system, *Applied Energy*, December 2015. DOI: 10.1016/j.apenergy.2015.12.035. 43

[81] C. Sun, X. Hu, S. J. Moura, and F. Sun, Velocity predictors for predictive energy management in hybrid electric vehicles, *IEEE Transactions on Control Systems Technology*, vol. 23, no. 3, pp. 1197–1204, May 2015. DOI: 10.1109/tcst.2014.2359176. 44, 51

[82] Q. Hu and W. Yue, *Markov Decision Processes with their Applications*, Springer-Verlag, NY, 2007. DOI: 10.1007/978-0-387-36951-8. 44

[83] L. Johannesson, M. Asbogard, and B. Egardt, Assessing the potential of predictive control for hybrid vehicle Powertrains using stochastic dynamic programming, *IEEE Transactions on Intelligent Transportation Systems*, vol. 8, no. 1, pp. 71–83, March 2007. DOI: 10.1109/itsc.2005.1520076. 44

[84] J. A. Gubner, *Probability and Random Processes for Electrical and Computer Engineers*, Cambridge University Press, NY, 2006. DOI: 10.1017/cbo9780511813610. 44

[85] S. J. Moura, H. K. Fathy, D. S. Callaway, and J. L. Stein, A stochastic optimal control approach for power management in plug-in hybrid electric vehicles, *IEEE Transactions on Control Systems Technology*, vol. 19, no. 3, pp. 545–555, May 2011. DOI: 10.1115/dscc2008-2252. 44

[86] C.-C. Lin, Modeling and Control Strategy Development for Hybrid Vehicles, Ph.D. dissertation, Dept. Mech. Eng., Univ. Michigan, Ann Arbor, MI, 2004. 44

[87] M. Josevski and D. Abel, Energy management of parallel hybrid electric vehicles based on stochastic model predictive control, *IFAC Proc.*, 47(3):2132–7, December 31, 2014. DOI: 10.3182/20140824-6-za-1003.01329. 45

[88] J. Wang, Y. Huang, H. Xie, and G. Tian, Driving pattern prediction model for hybrid electric buses based on real-world driving data, *KINTEX*, Korea, May 3–6, 2015. 45

[89] X. Zeng and J. Wang, A parallel hybrid electric vehicle energy management strategy using stochastic model predictive control with road grade preview, *IEEE Transactions on Control Systems Technology*, vol. 23, no. 6, pp. 2416–2423, November 2015. DOI: 10.1109/tcst.2015.2409235. 45

[90] L. Li, S. You, C. Yang, B. Yan, J. Song, and Z. Chen, Driving-behavior-aware stochastic model predictive control for plug-in hybrid electric buses, *Applied Energy*, vol. 162, pp. 868–879, January 2016. DOI: 10.1016/j.apenergy.2015.10.152. 45

[91] J. Zhang, H. He, X. Wang, Model predictive control based energy management strategy for a plug-in hybrid electric vehicle, *3rd International Conference on Mechanical Engineering and Intelligent Systems, ICMEIS*, 2015. DOI: 10.2991/icmeis-15.2015.165. 45

[92] T.-K. Lee, B. Adornato, and Z. S. Filipi, Synthesis of real-world driving cycles and their use for estimating PHEV energy consumption and charging opportunities: Case study for midwest/U.S., *IEEE Transactions on Vehicular Technology*, vol. 60, no. 9, pp. 4153–4163, November 2011. DOI: 10.1109/tvt.2011.2168251. 45

[93] F. Payri, C. Guardiola, B. Pla, and D. Blanco-Rodriguez, A stochastic method for the energy management in hybrid electric vehicles, *Control Engineering Practice*, vol. 29, pp. 257–265, August 2014. DOI: 10.1016/j.conengprac.2014.01.004. 45

[94] X. Zeng and J. Wang, A two-level stochastic approach to optimize the energy management strategy for fixed-route hybrid electric vehicles, *Mechatronics*, December 2015. DOI: 10.1016/j.mechatronics.2015.11.011. 46, 47

[95] A. D. Styler and I. R. Nourbakhsh, Model predictive control with uncertainty in human driven systems, *27th AAAI Conference on Artificial Intelligence*, June 2013. 46

[96] A. D. Styler and I. R. Nourbakhsh, Real-time predictive optimization for energy management in a hybrid electric vehicle, *AAAI*, pp. 737–744, January 25, 2015. 46

[97] M. T. Hagan, H. B. Demuth, M. H. Beale, and B. H. Demuth, *Neural Network Design*, Boston, PWS Pub., 1995. 46

[98] Y. Zhang, DIRECT Algorithm and Driving Cycle Recognition based Optimization Study for Hybrid Electric Vehicle, Ph.D. thesis, Dept. Mech. Eng., Beijing Inst. of Tech., Beijing, China, 2010. 46

[99] S. Jeon, S. Jo, Y. Park, and J. Lee, Multi-mode driving control of a parallel hybrid electric vehicle using driving pattern recognition, *Journal of Dynamic Systems, Measurement, and Control*, vol. 124, no. 1, p. 141, 2002. DOI: 10.1115/1.1434264. 46

[100] Y. Tian, X. Zhang, and L. Zhang, Intelligent energy management based on driving cycle identification using fuzzy neural network, *Proc. of the 2nd International Symposium on Computational Intelligence and Design*, Changsha, China, vol. 2, pp. 501–504, December 12–14, 2009. DOI: 10.1109/iscid.2009.271. 46

[101] R. Langari and J. S. Won, Integrated drive cycle analysis for fuzzy logic based energy management in hybrid vehicles, *Proc. of the 12th IEEE International Conference on Fuzzy Systems*, St. Louis, MI, pp. 290–295, May 25–28, 2003. DOI: 10.1109/fuzz.2003.1209377. 46

[102] H. He, C. Sun, and X. Zhang, A method for identification of driving patterns in hybrid electric vehicles based on a LVQ neural network, *Energies*, vol. 5, no. 12, pp. 3363–3380, September 2012. DOI: 10.3390/en5093363. 46

[103] I. Arsie, M. Graziosi, C. Pianese, G. Rizzo, and M. Sorrentino, Optimization of supervisory control strategy for parallel hybrid vehicle with provisional load estimate, *Proc. of AVEC04*, 23:23–7, August 2004. 46

[104] B. Y. Liaw and M. Dubarry, From driving cycle analysis to understanding battery performance in real-life electric hybrid vehicle operation, *Journal of Power Sources*, vol. 174, no. 1, pp. 76–88, November 2007. DOI: 10.1016/j.jpowsour.2007.06.010. 46

[105] M. Dubarry, V. Svoboda, R. Hwu, and B. Y. Liaw, A roadmap to understand battery performance in electric and hybrid vehicle operation, *Journal of Power Sources*, vol. 174, no. 2, pp. 366–372, December 2007. DOI: 10.1016/j.jpowsour.2007.06.237. 46

[106] Y. L. Murphey, et al., Intelligent hybrid vehicle power Control—Part II: Online intelligent energy management, *IEEE Transactions on Vehicular Technology*, vol. 62, no. 1, pp. 69–79, January 2013. DOI: 10.1109/tvt.2012.2217362. 46

[107] L. Niu, H. Yang, and Y. Zhang, Intelligent HEV fuzzy logic control strategy based on identification and prediction of drive cycle and driving trend, *World Journal of Engineering and Technology*, vol. 03, no. 03, pp. 215–226, 2015. DOI: 10.4236/wjet.2015.33c032. 46

[108] S. Zhang and R. Xiong, Adaptive energy management of a plug-in hybrid electric vehicle based on driving pattern recognition and dynamic programming, *Applied Energy*, vol. 155, pp. 68–78, October 2015. DOI: 10.1016/j.apenergy.2015.06.003. 47

[109] Q. Gong, Y. Li, Z. Peng, Power management of plug-in hybrid electric vehicles using neural network based trip modeling. *American Control Conference*, pp. 4601–4606, IEEE, June 10, 2009. DOI: 10.1109/acc.2009.5160623. 47

[110] Q. Gong and Y. Li, Trip based optimal power management of plug-in hybrid electric vehicle with advanced traffic flow modeling, *SAE*, 01–1316, 2008. DOI: 10.4271/2008-01-1316. 47

[111] M. Marx, X. Shen, and D. Soffker, A data-driven online identification and control optimization approach applied to a hybrid electric powertrain system, *IFAC Proc.*, 45(2):153–158, December 2012. DOI: 10.3182/20120215-3-at-3016.00027. 47

[112] Z. Chen, R. Xiong, and J. Cao, Particle swarm optimization-based optimal power management of plug-in hybrid electric vehicles considering uncertain driving conditions, *Energy*, vol. 96, pp. 197–208, February 2016. DOI: 10.1016/j.energy.2015.12.071. 47

[113] F. Soriano, M. Moreno-Eguilaz, and J. Alvarez-Florez, Drive cycle identification and energy demand estimation for refuse-collecting vehicles, *IEEE Transactions on Vehicular Technology*, vol. 64, no. 11, pp. 4965–4973, November 2015. DOI: 10.1109/tvt.2014.2382591. 47

[114] J. Unger, M. Kozek, and S. Jakubek, Nonlinear model predictive energy management controller with load and cycle prediction for non-road HEV, *Control Engineering Practice*, vol. 36, pp. 120–132, March 2015. DOI: 10.1016/j.conengprac.2014.12.001. 47

[115] C. H. Mayr, A. Fleck, and S. Jakubek, Hybrid powertrain control using optimization and cycle based predictive control algorithms, *9th IEEE International Conference on Control and Automation (ICCA)*, pp. 937–944, Santiago. http://dx.doi.org/10.1109/ICCA.2011.6138090 DOI: 10.1109/icca.2011.6138090. 47

[116] C. Sun, F. Sun, and H. He, Investigating adaptive-ECMS with velocity forecast ability for hybrid electric vehicles, *Applied Energy*, February 2016. DOI: 10.1016/j.apenergy.2016.02.026. 47

[117] T. Liu, Y. Zou, D. Liu, and F. Sun, Reinforcement learning of Adaptive energy management with transition probability for a hybrid electric tracked vehicle, *IEEE Transactions on Industrial Electronics*, vol. 62, no. 12, pp. 7837–7846, December 2015. DOI: 10.1109/tie.2015.2475419. 47

[118] Q. Gong, Y. Li, and Z.-R. Peng, Trip-based optimal power management of plug-in hybrid electric vehicles, *IEEE Transactions on Vehicular Technology*, vol. 57, no. 6, pp. 3393–3401, November 2008. DOI: 10.1109/tvt.2008.921622. 48

[119] C. Manzie, T. S. Kim, and R. Sharma, Optimal use of telemetry by parallel hybrid vehicles in urban driving, *Transportation Research Part C: Emerging Technologies*, vol. 25, pp. 134–151, December 2012. DOI: 10.1016/j.trc.2012.04.012.

[120] T. van Keulen, B. de Jager, A. Serrarens, and M. Steinbuch, Optimal energy management in hybrid electric trucks using route information, *Proc. Les Rencontres Scientifiques de l'IFP, Adv. Hybrid Powertrains*, Rueil-Malmaison, France, November 25–26, 2008. DOI: 10.2516/ogst/2009026. 48

[121] D. Ambuhl and L. Guzzella, Predictive reference signal generator for hybrid electric vehicles, *IEEE Transactions on Vehicular Technology*, vol. 58, no. 9, pp. 4730–4740, November 2009. DOI: 10.1109/tvt.2009.2027709. 48

[122] C. Zhang, A. Vahidi, P. Pisu, X. Li, and K. Tennant, Role of terrain preview in energy management of hybrid electric vehicles, *IEEE Transactions on Vehicular Technology*, 59(3):1139–1147, 2010. DOI: 10.1109/tvt.2009.2038707. 48

[123] L. Johannesson, S. Pettersson, and B. Egardt, Predictive energy management of a 4QT series-parallel hybrid electric bus, *Control Engineering Practice*, vol. 17, no. 12, pp. 1440–1453, December 2009. DOI: 10.1016/j.conengprac.2009.07.004. 48

[124] Y. He, J. Rios, M. Chowdhury, P. Pisu, and P. Bhavsar, Forward power-train energy management modeling for assessing benefits of integrating predictive traffic data into plug-in-hybrid electric vehicles, *Transportation Research Part D: Transport and Environment*, vol. 17, no. 3, pp. 201–207, May 2012. DOI: 10.1016/j.trd.2011.11.001. 48

[125] C. Sun, F. Sun, X. Hu, J. K. Hedrick, and S. Moura, Integrating traffic velocity data into predictive energy management of plug-in hybrid electric vehicles, *American Control Conference (ACC)*, pp. 3267–3272, IEEE, July 1, 2015 DOI: 10.1109/acc.2015.7171836. 48

[126] R. Bartholomaeus, M. Klingner, and M. Lehnert, Prediction of power demand for hybrid vehicles operating in fixed-route service, *Proc. of the 17th IFAC World Congress*, pp. 6–11, July 6, 2008. DOI: 10.3182/20080706-5-kr-1001.00951. 48

[127] M. A. Mohd Zulkefli, J. Zheng, Z. Sun, and H. X. Liu, Hybrid powertrain optimization with trajectory prediction based on inter-vehicle-communication and vehicle-infrastructure-integration, *Transportation Research Part C: Emerging Technologies*, vol. 45, pp. 41–63, August 2014. DOI: 10.1016/j.trc.2014.04.011. 49

[128] T. Van Keulen, B. de Jager, D. Foster, and M. Steinbuch, Velocity trajectory optimization in hybrid electric trucks, *Proc. of the American Control Conference*, pp. 5074–5079, IEEE, June 30, 2010. DOI: 10.1109/acc.2010.5530695. 49

[129] Fu L., Ümit Ö, P. Tulpule, V. Marano, Real-time energy management and sensitivity study for hybrid electric vehicles, *Proc. of the American Control Conference*, pp. 2113–2118, IEEE, June 29, 2011. DOI: 10.1109/acc.2011.5991374. 49

[130] L. Johannesson, M. Nilsson, and N. Murgovski, Look-ahead vehicle energy management with traffic predictions, *IFAC-PapersOnLine*, 48(15):244–51, December 31, 2015. DOI: 10.1016/j.ifacol.2015.10.035. 49

[131] P. G. Gipps, A behavioural car-following model for computer simulation, *Transportation Research Part B: Methodological*, vol. 15, no. 2, pp. 105–111, April 1981. DOI: 10.1016/0191-2615(81)90037-0. 49

[132] F. Jiménez and W. Cabrera-Montiel, System for road vehicle energy optimization using real time road and traffic information, *Energies*, vol. 7, no. 6, pp. 3576–3598, June 2014. DOI: 10.3390/en7063576. 49

[133] K. Yu, et al., Model predictive control for connected hybrid electric vehicles, *Mathematical Problems in Engineering*, pp. 1–15, 2015. DOI: 10.1155/2015/318025. 49

[134] L. Johannesson, N. Murgovski, E. Jonasson, J. Hellgren, and B. Egardt, Predictive energy management of hybrid long-haul trucks, *Control Engineering Practice*, vol. 41, pp. 83–97, August 2015. DOI: 10.1016/j.conengprac.2015.04.014. 49

[135] C. Sun, S. J. Moura, X. Hu, J. K. Hedrick, and F. Sun, Dynamic traffic feedback data enabled energy management in plug-in hybrid electric vehicles, *IEEE Transactions on Control Systems Technology*, vol. 23, no. 3, pp. 1075–1086, May 2015. DOI: 10.1109/tcst.2014.2361294. 49

[136] G. E. Katsargyri, I. V. Kolmanovsky, J. Michelini, M. L. Kuang, A. M. Phillips, M. Rinehart, and M. A. Dahleh, Optimally controlling hybrid electric vehicles using path forecasting. *Institute of Electrical and Electronics Engineers*. DOI: 10.1109/acc.2009.5160504. 49

[137] X. Li, Z. Sun, D. Cao, Z. He, and Q. Zhu, Real-time trajectory planning for autonomous urban driving: Framework, algorithms, and verifications, *IEEE/ASME Transactions on Mechatronics*, vol. 21, no. 2, pp. 740–753, April 2016. DOI: 10.1109/tmech.2015.2493980. 50

[138] J. Ji, A. Khajepour, W. Melek, and Y. Huang, Path planning and tracking for vehicle collision avoidance based on model predictive control with multi-constraints, *IEEE Transactions on Vehicular Technology*, pp. 1–1, 2016. DOI: 10.1109/tvt.2016.2555853. 50

[139] A. Fotouhi, R. Yusof, R. Rahmani, S. Mekhilef, and N. Shateri, A review on the applications of driving data and traffic information for vehicles' energy conservation, *Renewable and Sustainable Energy Reviews*, vol. 37, pp. 822–833, September 2014. DOI: 10.1016/j.rser.2014.05.077. 50

[140] S. Kermani, S. Delprat, T. M. Guerra, R. Trigui, and B. Jeanneret, Predictive energy management for hybrid vehicle, *Control Engineering Practice*, vol. 20, no. 4, pp. 408–420, April 2012. DOI: 10.1016/j.conengprac.2011.12.001. 51

[141] R. Beck, A. Bollig, and D. Abel, Comparison of two real-time predictive strategies for the optimal energy management of a hybrid electric vehicle, *Oil and Gas Science and Technology—Revue de l'IFP*, vol. 62, no. 4, pp. 635–643, July 2007. DOI: 10.2516/ogst:2007038. 51

[142] C. C. Lin, H. Peng, S. Jeon, and J. M. Lee, Control of a hybrid electric truck based on driving pattern recognition, *Proc. of the Advanced Vehicle Control Conference*, Hiroshima, Japan, September 2002. 51

[143] E. Tazelaar, J. Bruinsma, B. Veenhuizen, and P. van den Bosch, Driving cycle characterization and generation, for design and control of fuel cell buses, *World Electric Vehicle Journal*, 3:1–8, May 2009. 51

[144] D. Shen, V. Bensch, and S. Miiller, Model predictive energy management for a range extender hybrid vehicle using map information, *IFAC-Papers OnLine*, 48(15):263–70, December 31, 2015. DOI: 10.1016/j.ifacol.2015.10.038. 51

[145] R. Langari and J. Won, Intelligent energy management agent for a parallel hybrid vehicle—Part I: System architecture and design of the driving situation identification process, *IEEE Transactions on Vehicular Technology*, vol. 54, no. 3, pp. 925–934, May 2005. DOI: 10.1109/tvt.2005.844685. 51

[146] K. Igarashi, C. Miyajima, et al., Biometric identification using driving behavioral signals, *IEEE International Conference on Multimedia and Expo*, vol. 1, pp. 65–68, June 2004. DOI: 10.1109/icme.2004.1394126. 51

[147] H. Wang, et al., A novel energy management for hybrid off-road vehicles without future driving cycles as a priori, *Energy*, 2017. DOI: 10.1016/j.energy.2017.05.172. 50, 67

[148] M. Pan, et al., Fuzzy control and wavelet transform-based energy management strategy design of a hybrid tracked bulldozer, *Journal of Intelligent and Fuzzy Systems*, Preprint, pp. 1–10. DOI: 10.3233/ifs-151959. 52, 61

[149] Q. Xiao, Q. Wang, and Y. Zhang, Control strategies of power system in hybrid hydraulic excavator, *Automation in Construction*, vol. 17, no. 4, pp. 361–367, May 2008. DOI: 10.1016/j.autcon.2007.05.014. 52

[150] X. Lin, S. Pan, and D. Wang, Dynamic simulation and optimal control strategy for a parallel hybrid hydraulic excavator, *Journal of Zhejiang University-SCIENCE A*, vol. 9, no. 5, pp. 624–632, May 2008. DOI: 10.1631/jzus.a071552. 53

[151] P. Roskilly, R. Palacin, and J. Yan, Novel technologies and strategies for clean transport systems, *Applied Energy*, vol. 157, pp. 563–566, November 2015. DOI: 10.1016/j.apenergy.2015.09.051. 65

[152] Y. Huang, H. Wang, A. Khajepour, et al., Model predictive control power management strategies for HEVs: A review [J]. *Journal of Power Sources*, 341:91–106, 2017. DOI: 10.1016/j.jpowsour.2016.11.106. 65

[153] H. Borhan and A. Vahidi, Model predictive control of a hybrid electric powertrain with combined battery and ultracapacitor energy storage system, *International Journal of Powertrains*, vol. 1, no. 4, p. 351, 2012. DOI: 10.1504/ijpt.2012.049645. 66, 67

[154] E. Schaltz, A. Khaligh, and P. O. Rasmussen, Influence of battery/Ultracapacitor energy-storage sizing on battery lifetime in a fuel cell hybrid electric vehicle, *IEEE Transactions on Vehicular Technology*, vol. 58, no. 8, pp. 3882–3891, October 2009. DOI: 10.1109/tvt.2009.2027909. 66, 67

[155] N. Omar, J. Van Mierlo, B. Verbrugge, and P. Van den Bossche, Power and life enhancement of battery-electrical double layer capacitor for hybrid electric and charge-depleting plug-in vehicle applications, *Electrochimica Acta*, vol. 55, no. 25, pp. 7524–7531, October 2010. DOI: 10.1016/j.electacta.2010.03.039. 66, 67

[156] M. Masih-Tehrani, M.-R. Haíri-Yazdi, V. Esfahanian, and A. Safaei, Optimum sizing and optimum energy management of a hybrid energy storage system for lithium battery life improvement, *Journal of Power Sources*, vol. 244, pp. 2–10, December 2013. DOI: 10.1016/j.jpowsour.2013.04.154. 67

[157] M. Ortuzar, J. Moreno, and J. Dixon, Ultra capacitor-based auxiliary energy system for an electric vehicle: Implementation and evaluation, *IEEE Transactions on Industrial Electronics*, vol. 54, no. 4, pp. 2147–2156, August 2007. DOI: 10.1109/tie.2007.894713. 67

[158] J. M. Miller, P. J. McCleer, and M. Everett, Comparative assessment of ultra-capacitors and advanced battery energy storage systems in PowerSplit electronic-CVT vehicle powertrains, *IEEE International Conference on Electric Machines and Drives*, IEEE, 2005. DOI: 10.1109/iemdc.2005.195921. 70

[159] O. Onar and A. Khaligh, Dynamic modeling and control of a cascaded active battery/ultra-capacitor based vehicular power system, *IEEE Vehicle Power and Propulsion Conference*, 2008. DOI: 10.1109/vppc.2008.4677598. 70

[160] V. Marano, et al., Lithium-ion batteries life estimation for plug-in hybrid electric vehicles, *IEEE Vehicle Power and Propulsion Conference*, 2009. DOI: 10.1109/vppc.2009.5289803. 69, 70

[161] L. Serrao, et al., Optimal energy management of hybrid electric vehicles including battery aging, *Proc. of the American Control Conference*, IEEE, 2011. DOI: 10.1109/acc.2011.5991576. 70, 71

[162] L. Serrao, et al., A novel model-based algorithm for battery prognosis, *IFAC Proc.*, vol. 42.8, pp. 923–928, 2009. DOI: 10.3182/20090630-4-es-2003.00152. 71

[163] V. Marano and G. Rizzoni, Energy and economic evaluation of PHEVs and their interaction with renewable energy sources and the power grid, *Vehicular Electronics and Safety, ICVES, IEEE International Conference on*, 2008. DOI: 10.1109/icves.2008.4640909. 71

[164] H. J. Ferreau, C. Kirches, A. Potschka, et al., qpOASES: A parametric active-set algorithm for quadratic programming, *Mathematical Programming Computation*, 6(4):327–363, 2014. DOI: 10.1007/s12532-014-0071-1. 62, 73

[165] M. G. Bekker, *Theory of Land Locomotion*, The Univ. of Michigan Press, Ann Arbor, MI, 1956. DOI: 10.3998/mpub.9690401. 5, 7, 12

Authors' Biographies

HONG WANG

Hong Wang is currently a Research Associate of Mechanical and Mechatronics Engineering with the University of Waterloo. She received her Ph.D. from the Beijing Institute of Technology in China in 2015. Her research focuses on the component sizing, modeling of hybrid powertrains, and energy management control strategies design for hybrid electric vehicles; intelligent control theory and application; and autonomous vehicles.

YANJUN HUANG

Yanjun Huang is currently a Postdoctoral Fellow of Mechanical and Mechatronics Engineering with the University of Waterloo, where he received his Ph.D. in 2016. He received an M.S. degree in vehicle engineering from Jilin University, China in 2012. He is working on advanced control strategies and their real-time applications; vehicle dynamics and control; autonomous vehicle; Heating, Ventilating, and Air Conditioning (HVAC) system modeling and control; modeling of hybrid powertrains, components sizing, and energy management control strategies design through concurrent optimization and HIL testing; and variable valve actuation system for engines.

AMIR KHAJEPOUR

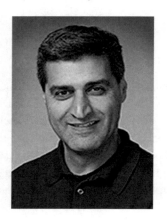

Amir Khajepour is a professor in the department of Mechanical and Mechatronics Engineering at the University of Waterloo. He holds the Canada Research Chair in Mechatronic Vehicle Systems, and NSERC/General Motors Industrial Research program that applies his expertise in several key multidisciplinary areas including system modeling and control of dynamic systems. His research has resulted in many patents and technology transfers. He is the author of more than 400 journal and conference publications as well as several books. He is a Fellow of the Engineering Institute of Canada, the American Society of Mechanical Engineers, and the Canadian Society of Mechanical Engineering.

CHUAN HU

Chuan Hu received a B.E. degree in vehicle engineering from Tsinghua University, Beijing, China, in 2010; an M.E. degree in vehicle operation engineering from the China Academy of Railway Sciences, Beijing, China, in 2013; and a Ph.D. degree in Mechanical Engineering from McMaster University, Hamilton, Canada, in 2017. He is now a Postdoctoral Fellow with the Department of Systems Design Engineering, at the University of Waterloo, Waterloo, Canada. His research interests include vehicle system dynamics and control, motion control and estimations of autonomous vehicles, mechatronics, and robust and adaptive control.

Printed in the United States
by Baker & Taylor Publisher Services